The Architect's Engineer

The Architect's Engineer

*Memoir of a Life in Building
Engineering and Restorative Development*

Sir Nigel Thompson

Foreword by Sir Terry Farrell

McFarland & Company, Inc., Publishers
Jefferson, North Carolina

Frontispiece: Portrait of Nigel Thompson by painter, illustrator and designer Peter M.J. Lightfoot for the D Group in February 1996. The D Group, a Strategic Business Development Company, was founded by Paul Cautley in 1994, and Nigel Thompson was one of the founder members (portrait copyright Peter M.J. Lightfoot, Dorchester, Dorset).

ISBN (print) 978-1-4766-8589-2
ISBN (ebook) 978-1-4766-4524-7

LIBRARY OF CONGRESS AND BRITISH LIBRARY
CATALOGUING DATA ARE AVAILABLE

Library of Congress Control Number 2022019992

© 2022 Nigel Thompson. All rights reserved

No part of this book may be reproduced or transmitted in any form or by any means, electronic or mechanical, including photocopying or recording, or by any information storage and retrieval system, without permission in writing from the publisher.

Front cover: Photograph of Nigel Thompson aged 60, when he was asked by U.K. Prime Minister Tony Blair to form and lead a U.K. Task Force from the Public and Private Sectors for the Reconstruction of Kosovo; background: London panorama with Embankment Place by Terry Farrell, Hungerford Bridge over Thames River (Charing Cross Bridge), Victoria Embankment and urban architectures (Shutterstock/Premier Photo)

Printed in the United States of America

McFarland & Company, Inc., Publishers
Box 611, Jefferson, North Carolina 28640
www.mcfarlandpub.com

For Nicky, Nim, Sam and Phoebe,
Taia, Milly and Tom

My humanity is bound up in yours, for we can only be human together.

—Archbishop Desmond Tutu

Table of Contents

Acknowledgments — ix
Foreword by Terry Farrell — 1
Preface — 3

Part I—An Aspiring Building Engineer   5

1. London During the War (1939–1945) — 7
2. Willington School and St. Paul's School (1945–1955) — 10
3. Starting Work and Rhodesia (1955–1960) — 17
4. Return to England—Scholarship and Joining Arup (1960) — 26
5. Friends and London in the Early 1960s — 32
6. Meeting Nicky (1963) — 34
7. Summer in Turkey (1964) — 36
8. Senior Engineer and Marrying Nicky (1965) — 38
9. The Mill and Nim's Arrival (1966) — 43
10. Thompson's Eating House (1970–1972) — 46
11. Grove Farm and Sam's Arrival (1972) — 49
12. Iran and Phoebe's Arrival (1974–1976) — 53
13. Doha and Kafrawi's Tower of Winds (1976–1978) — 60
14. Life in Wiltshire and Africa Again (1977) — 66
15. Commercial Buildings in the 1980s — 72
16. Theaters and The Old Vic (1982) — 75

Part II—A Leader in Building Engineering and Business Development   79

17. Joining the Arup Main Board (1984) — 81
18. British Overseas Trade Board and the Vineyard (1983–1984) — 85

19. Hospitals in the 1980s: St. Mary's, London, UK; Onassis Hospital, Athens, Greece; Yangon Hospital, Burma (Myanmar) — 88
20. Shopping Centers and Retail Developments in the 1980s — 93
21. Lutyens House (1986–1987) — 97
22. Nicky's Singing — 102
23. Air Rights Buildings (1985) and Working with Terry Farrell — 105
24. The End of the Vineyard and Arab Horse Racing (1988) — 116
25. Fiftieth Birthday and Twenty-Fifth Wedding Anniversary (1989) — 119
26. Kuwait City (1991) — 124
27. Nicky's Accident — 129
28. Business Development (1992) — 132
29. Zimbabwe (1997) — 141
30. Financial Services Sector (1990s) — 145

Part III—A Leader in Restorative Development 149

31. Power in Kosovo — 151
32. Serbia and Montenegro (2000) — 166
33. Stability Pact for South-Eastern Europe (2001) — 173
34. The Island of St. Helena (2002) — 178
35. Campaign to Protect Rural England (2003) — 188
36. The Marlborough Brandt Group and BUILD (2004) — 194
37. Action for the River Kennet (ARK) — 201
38. Governor of St. Paul's School (2007–2011) and President of Old Boys (2015–2017) — 204
39. Minal Parish Council (2000–2018) — 210
40. The Trevor Estate (1976–2011) — 212
41. Grove House and Estate — 214

Appendix I: Nigel Thompson Selected Titles and Awards — 225
Appendix II: Terms and Abbreviations — 226
Chapter Notes — 229
Bibliography — 249
Index — 251

Acknowledgments

This book would not exist without my friend John Pascoe who had the idea of writing my biography after hearing of my achievements in Kosovo. He did a lot of research without any encouragement from me. Then the project was dormant until 2020, when my wife Nicky persuaded me to write my autobiography so that our family would hear about my life over the last 80 years. I have to thank Nicky for her support and patience (an attribute that she is not known for!).

I would like to thank my friend Sir Terry Farrell for writing the Foreword for my book, General Sir Mike Jackson and Marc Franco for recording my work in Kosovo, Tim Haig for writing on our promotional work together and, particularly, I want to thank Anne Kriken Mann for reading through my manuscript and for her considerable support and encouragement.

I would also like to thank my daughter Nim and my grandson Tom for collating and sometimes accidently eliminating family photographs on my computer, Peter M.J. Lightfoot for permitting use of his portrait of me as the book's Frontispiece, Danny Imade for all his work in selecting suitable photographs from the Arup Photolibrary, Tim Cunis and Michael Knott for finding in school archives a photograph of me at St. Paul's School in 1952–53, Peter Allport and Colin Weyer of www.rhodesia.me.uk for brilliant photographs covering my time in Rhodesia, Michael and Carole Taylor and Malcolm and Erica Barrie in finding photographs on Arup projects, Brian Stone of Mott MacDonald for additional photographs on Kosovo, Anna Forbes and Helen Llewelyn at elcotcreative.co.uk for providing pictures of the river Kennet and ARK activities, Julian Diamond for support and information on Arup, Marc Prior, Jordan and Michal Witkowski for keeping my computer working and Sue Hine for typing my first draft.

The last and greatest of my debts go to Nicky, not only for encouraging me in the production of this book but for being with me for the last 55 years of our married life.

Foreword
by Terry Farrell

I first met Nigel Thompson in the very early 1980s. This was not long after I set up on my own practice, having been in partnership with Nick Grimshaw for 15 years. I was well used to high-tech collaborations between architects and engineers, where an architect glorified engineering for its own sake, and to my mind the engineer hid behind the architect who stole his identity somewhat.

But by the early 1980s I had arranged to meet three of the teams at Arup because I was looking for a new way forward. I found in Nigel what I had been looking for. He immediately struck me as a more well-balanced engineer. He was more independent and responded to a more rounded view of context, history and city-making. He was not a specialist looking for technical wizardry per se, that would impress by preconceived notions of style. He and I got on instinctively and our practices grew and grew. He always joined with me in making each project the best of the best.

The early projects were mainly competitions, but in the late 1980s, two projects stood out: Embankment Place (above Charing Cross Station) and Alban Gate (above the junction of Wood Street and London Wall in the City of London). They were both air rights buildings, and the engineering was critical to the design—but in a way that added to the architecture in a collaborative way that spoke volumes about the excellent understanding that existed between us. For example, at Embankment Place, I drew from the outset an expressive roofscape as history and context I felt demanded it. On hearing this, Nigel's team came up with the idea of capitalizing on these ambitions by adopting pronounced and expressed arched structures that had all the floors suspended from them. It was a brilliant engineering concept that without my sketches might not have occurred to Nigel's team.

This would be the way we worked together to solve the fire escape staircases placed outside the floorplans of each of the four towers topped by individual meeting rooms. Also, I remember the lifts core—searching for a central space until together we found that a central cab rank was ideally positioned to fulfill this function, and so it went on. The acoustic solution was a waterfall to drown the sounds of shunting trains above the main entrance hall (all researched by Arup Acoustics). I remember well the 18 enormous piles dug below station level, because we were "gifted" the tracks, which were suspended above the arches, leaving ample space below to do the digging without interfering with the running of the trains above. I say digging as there was not enough space for a piling rig which is why, most unusually, the piles were hand dug by miners who were available because of the strikes at home in Yorkshire. The piles were 35 meters deep with a huge under-ream in excess of six meters in diameter. It was like a large room, and when each one was finished, the miners held a "bottoming out" party. They worked at terrific speed and, like the entire engineering of this project, were a great success.

Alban Gate was a different type of air rights building—being more vertical and therefore having more expressive potential on the façades. Nigel's team's solution was, in order to provide additional stiffness as the front tower was above roads and therefore lacked rigid grounding, to express large atriums facing St. Paul's with exposed structure, differently composed by us in each four-story element. It was really a triumph of engineering and architecture working together without one dominating the other in a subservient role.

Our collaborations extended to master planning in the late 1980s and early 1990s. Two examples of Nigel's thoughtful solutions stand out. The first was a masterplan study for the North West Hospital Board, who were grappling with University College and Westminster Hospitals' future. They said at the interview before lunch that we lacked a hospital planning specialist. After a good lunch nearby, Nigel and I were motivated to write to the panel and say we had thought of a hospital consultant and he had joined us. By that evening we had won the job. The other was Edinburgh Exchange after a heavily loaded infrastructure began to give us problems due to the recession of the early 1990s. He and I went to Edinburgh Council and radically simplified the infrastructure, convincing them to proceed with the project. Both masterplans were successfully built out.

After we opened an office in Hong Kong, Arup provided an office space and help to get us established. However, Arup grew exponentially in the 1990s. We continued to do work with them, but I lost touch with Nigel as he moved to the upper echelons of the organization. His wife's accident affected all of us. Nigel concentrated instead on Kosovo and St. Helena, reinstatements and holistic thinking projects which he was highly suited to.

I came across his professional career again when he was head of Campaign to Protect Rural England, which is a cause and sector I strongly identify with. He exhibited the strong leadership (almost modest to a fault) and sensitive thoughtfulness that I always associated with him. On a personal side I have always counted him as a true friend, as he is "what it says on the tin," a genuine, honorable and thoughtful person. He epitomizes 50 years of post-war leadership in creative placemaking, and I'm proud of my part in his professional life.

Terry Farrell—September 2020

Sir Terry Farrell CBE is considered to be one of the UK's leading architects and urban designers. Sir Terry was named CBE in 1996 and made a Knight Bachelor in 2001. In 2013 Sir Terry was voted the individual who made the Greatest Contribution to London's Planning and Development over the last ten years. In 2017, Sir Terry was awarded the Royal Town Planning Institute's Gold Medal in recognition of his outstanding achievements as one of the world's most influential architects, planners and urban designers.

Preface

I grew up in the London Blitz, going to school around the craters and rubble. It made me want to build things up rather than knock them down. As a child I was fascinated by the Anderson and Morrison bomb shelters which I had helped my father erect. This was my first adventure in building engineering. There was a photograph of my father and I building the shelter, and it would be a cherished possession if I still had it ... but it is many years disappeared. Losing my father aged eight increased my desire to become an engineer. The simple effectiveness of the shelter's design influenced my approach when I began designing structures.

After working in Africa I was fortunate to join Arup because it was one of very few firms at that time to have a philosophy. I enthusiastically followed the teachings of Ove Arup and the charismatic, dynamic leadership of the firm's CEO, Peter Dunican (PD). I was offered a place on the 60-strong Sydney Opera House team but asked for something smaller to start on. So PD assigned me to the National Recreation Centre team at Crystal Palace, where I worked with Poul Beckmann—a brilliant engineer and mentor.

When I joined Arup in 1960, the firm offered structural engineering with fewer than 300 employees working through offices in four countries (UK, Ireland, Nigeria, South Africa). Today Arup offers more than 90 disciplines with more than 15,000 employees working through offices in more than 80 offices worldwide.

I loved being part of that transformation, leading a multidisciplinary building engineering team designing amazing buildings and particularly those with the brilliant architect Terry Farrell. The highlight of my work at Arup was undoubtedly designing the air rights buildings with Terry Farrell. At Embankment Place we built a huge new office block over Charing Cross Railway Station, all hung from nine bow-string arches. At Alban Gate we built a second air rights building spanning over the intersection of London Wall and Wood Street.

I was then additionally taking responsibility for the firm's business development globally and helping to introduce into the firm new skills such as fire safety design and computational fluid dynamics (CFD). But I stepped down from the Arup main board in the spring of 1999. That was because I wanted to spend more time with my wife Nicky, who'd had a horse riding accident and—having lost her active lifestyle—was finding being disabled extremely frustrating. Arup made me

Non-Executive Deputy Chairman of Ove Arup & Partners so that I could continue with Business Development for the firm.

A few months later Slobodan Milosovic drove the Albanian Kosovans out of their homes in an act of ethnic cleansing. The United Nations agreed to a bombing campaign, followed by an invasion by land forces to drive out the Serbian Army and enable the displaced local communities to return. I was asked by Tony Blair, UK Prime Minister, to form and lead a UK task force from the public and private sectors for the reconstruction of Kosovo. This was my first post–Arup incursion into what is now termed "restorative development."

Around the time of the successful return of the Albanian Kosovans, John Pascoe, a colleague at Arup, approached me with the idea of doing an autobiography. John did research and drafting, and we did a few interviews. But I just didn't get time to progress the book. Then in March 2020 Covid-19 locked me down. Nicky, supported by the family, encouraged me to take up my pen and write so there would be a record of my life from London during the war, through working in Africa, joining Ove Arup & Partners and my worldwide experiences, particularly in Iran and the Arabian Gulf. The resulting narrative interweaves my family life with my worldwide experience working as an engineer.

So what made me so busy that I couldn't progress the book between 2003 and 2020? I was working on projects of a distinctively restorative development nature. I worked in Serbia and Eastern Europe on building rehabilitation, urban regeneration, the environment, ecosystems, water resources, power supplies, transportation and infrastructure. Much of what I was now doing was based around sustainability, conservation and quality of life. Most notably, I was fortunate to be involved in some countryside and community issues with my friends and CPRE (Campaign to Protect Rural England) colleagues Max Hastings and Bill Bryson.

Only a select few of my hundreds of building projects could be featured in this book. Authoritative and copiously illustrated technical papers on these select few, extraordinary and exciting projects are listed in the Bibliography. I've included at the foot of the Bibliography an Arup web address where readers of this book can read the technical papers free of charge.

I'm delighted that eminent North American academic publisher McFarland of Jefferson, North Carolina, has published my autobiography. A consequence is that readers in the United Kingdom will notice that the spelling throughout is in American English and not my usual Queen's English.

Part I

An Aspiring Building Engineer

"We are led to seek overall quality, fitness for purpose, as well as satisfying, or significant, forms and economy of construction. To this must be added harmony with the surroundings and the overall plan. We are then led to the ideal of *Total Architecture*, in collaboration with other like-minded firms or, better still, on our own. This means expanding our field of activity into adjoining fields: architecture, planning, ground engineering, environmental engineering, computer programming, etc. and the planning and organisation of the work on site. It is not the wish to expand, but the quest for quality which has brought us to this position, for we have realised that only intimate integration of the various parts or the various disciplines will produce the desired result."

—Ove Arup (1895–1988)

"Today, structural engineers are concerned mainly with the viability, stability, efficiency and economy of their constructions, or systems, to use a general expression, that they are creating. They are not necessarily concerned with the physical environment of the systems except where the environment affects the system. But it is the effect of the system on the environment that should now concern and obsess us. The aesthetic consequences of our calling are tomorrow's imperative and can be realised satisfactorily only through the total and willing collaboration of all concerned. It is no longer a matter of skill or competence or controls or regulations. It is primarily a question of our attitude and willingness to trust and to work together with our collaborators—as we want them to trust and to work with us—to achieve our collective aims. This is what creating the built environment is all about."

—Peter Dunican (1918–1989)

Part I—An Aspiring Building Engineer

Nigel Cooper Thompson—A Life Story

Born June 18, 1939, at No. 13, Temple Sheen, London SW14
Father: Henry Cooper Thompson 1/27/1908–5/23/1948
Mother: Beatrix Mary Cooper Thompson 12/27/1911–April 1972
Brother: Peter Cooper Thompson 5/21/1937–12/12/1991

My father Henry Cooper Thompson and my mother Beatrix Mary Cooper Thompson (née Henderson) on their wedding day (author's collection).

1

London During the War (1939–1945)

    I was born in London three months before the commencement of World War II. We spent the whole war in London. I vividly remember the nights of bombing, hearing the air raid warning and running to the underground "Anderson" shelter,[1] which I had helped my father build at the bottom of the garden—the dog was always first into the shelter! As we ran the sky was filled with the crisscross beams of the search lights, and there were flashes as the ack-ack guns[2] fired at the bombers. We sat in the shelter while the ground shook, mostly staying in there all night. After a particularly large explosion my father would gently open the door to see if the house was still there.

    I remember coming out in the mornings and, with my parents, inspecting the house for damage—there were invariably lumps of very hot shrapnel lying around, broken windows, and, on one occasion, my mother burst into tears when she saw one of her glass vases smashed to pieces from flying shrapnel. Similar scenes were being enacted in urban areas throughout the country. Some 3.5 million Anderson shelters were built, and each one came with directions for assembly and erection which included the parts diagram reproduced in this chapter.

    During the war my father was working as an engineer for the Ministry of Supply. He had wanted to join up and fly for the Royal Air Force as he already had a pilot's license, but, according to my mother, he was considered too valuable designing tanks at the Ministry. At nights my father would be out working with the ARP (Air Raid Precautions) pulling people out of bombed buildings. Sometimes when he returned in the early hours—we were under the stairs or in the Anderson shelter—he would be covered with blood and dust.

    We were all very upset when we heard that my uncle, mother's brother Ian Henderson, was killed. He was a Royal Navy Captain of a motor launch which took on German destroyers to enable the rest of the British fleet to escape Saint-Nazaire where they had carried out a raid to destroy the port. Their whole ship was awarded the Victoria Cross for outstanding bravery.

    Peter and I went to school, avoiding the bomb craters and rubble, and during air raids we would all go into the school cellar. This was Willington School in Putney. Later in the war Peter and I would sleep, at night, in the "Morrison" shelter[3] under the dining room table. The "Morrison" was the improved household shelter comprising a heavy steel box 6 ft. (1.8 m.) by 4 ft. (1.2 m.) and 3 ft. (0.9 m.) deep. The Anderson referred to earlier was a corrugated steel bunker at the bottom of the garden. I was fascinated by both structures.

I remember my first day at the pre-prep school—aged three. It was horrific because it was a girls' convent, which took boys and girls up to about eight years old, and the noise of large girls shrieking or wanting to kiss me is my only memory.

My next recollection is, aged five, starting at my prep school in Putney. It was a huge improvement on the nursery school as it was only for boys! I was good at maths and could not understand why other boys found sums so difficult, but I was slow at reading. I also had to spend a long time in the dining room because I refused to eat my cabbage, Brussels sprouts or swede. I would just sit there being obstinate until I learned to keep them in my cheek for removal later.

Other recollections from the War include seeing the farmers harvesting the corn in Richmond Park and shooting the rabbits as the cover diminished. Also I recall keeping chickens and pet rabbits, which we subsequently ate. I remember on one occasion when we had my pet rabbit "Thumper" for Sunday lunch—my parents weren't hungry!

I remember my parents were understandably very upset when my father had to shoot our lovely Alsatian dog (because he was very ill). But they were even more upset by my asking when were we going to eat him.

I remember standing on the lawn with a whistle watching for "doodlebugs"[4]—the German flying bomb which fell when the engine cut out. There was, apparently, just time if the engine cut out for us to run for cover! Or maybe it was mother keeping me occupied responsibly! Of course, I remember the great party at the end of the war—Victory in Europe (VE) day. All

My mother, my brother Peter and me (standing on the chair) (author's collection).

the lights in the street of every house were on. We had a huge bonfire in the garden. Everyone was laughing and dancing; everyone was happy. We didn't have to go to bed! It was great. We had been very lucky because, throughout the War, milk, vegetables and bread were delivered to our house in East Sheen by horse and cart, and we had all survived.

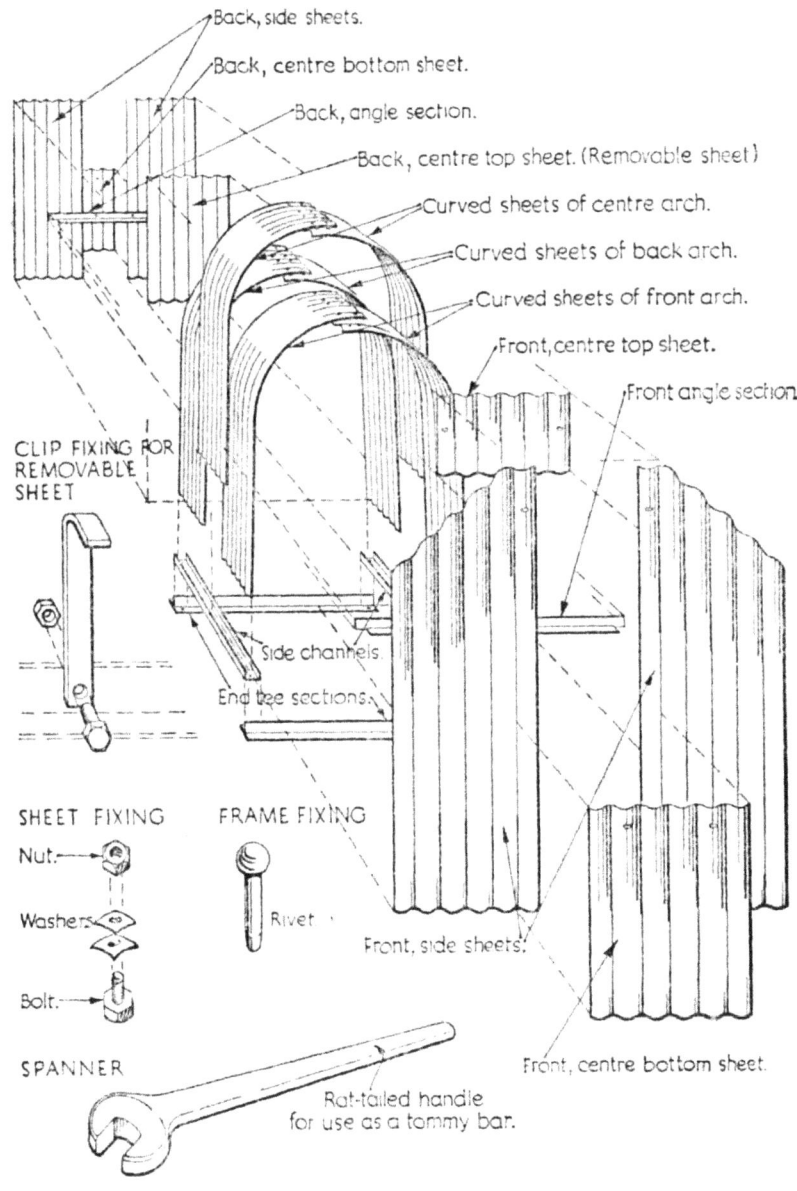

Individual Parts of the Anderson Shelter (Home Office, Air Raid Precautions: Directions for the Erection and Sinking of the Galvanised Corrugated Steel Shelter, Crown Copyright, February 1939).

2

Willington School and St. Paul's School (1945–1955)

This was my time at Willington School.[1] Among 120 boys aged from 5 to 13, I was good at maths and art but average at everything else. After six years I was in the first rugby, cricket and football teams. I was good but not a star player. However, I was very good at rifle shooting!

My father died in 1948 when I was eight, nearly nine. I missed him very much, and I missed having a father. Although I knew my mother tried hard to compensate, even coming to watch me play cricket on a few occasions, I envied my friends who had a father to help make things and to play cricket and other games.

Special occasions with my father included when, at the end of the war, he bought an old MTB (motor torpedo boat). It was great fun staying on the boat. He planned to convert it into a light motor cruiser. It was first moored at Itchenor (near Chichester) and then at Looe in Cornwall. Peter and I learned to row dinghies and generally mucked about at being naval captains and firing torpedoes. I don't think mother enjoyed our two holidays on the boat.

I remember also going to watch Father play cricket at Rosslyn Park. He was a good fast bowler.

I remember once having an argument with my father. It was on our boat, and it was over lunch. I refused to drink my green pea soup because I thought it was thick, lumpy and horrid. The row ended with my father smearing the soup all over my face! I am told that in a fury I said I would break all my parents' china and steal all their silver!

My father died of leukemia when he was 40. He had been very ill for some time—in and out of hospital. When I went into my mother's bedroom to say "good morning," she told Peter and me that our father had died during the night. We sat by the gas fire in her bedroom and cried.

Peter was sent to stay with Grannie Henderson and mother's two sisters, Oona and Yvonne, at their farmhouse in Bracknell. I went to stay with Grandma and Grandpa Thompson in Norwich. We stayed away from Mother for a week or two while she had to organize my father's cremation at Mortlake crematorium.

During my stay in Norfolk, I went sailing on the Norfolk Broads in an old

2. Willington School and St. Paul's School

Sports day at Willington School, Putney, London SW15. I am number 69 (white arrow). The whistle to start the race is about to be blown by Mr. Crebbin (author's collection).

Thames barge—great fun. We went to Lowestoft and visited some of my father's cousins. My grandfather was William Alfred Thompson. He was a board director of Norwich Building Society[2] and, I think, a Norwich city councillor. He was the youngest of a family of 14. The family owned a steelworks in Norfolk. He married Ada Blanche Cooper, who was an only child. Hence their children were named Cooper-Thompson. They had three children: my father Henry, the eldest, then Molly, and finally Ralph. Molly married Harold Butcher, whose family had a large haberdasher's[3] store in the center of Norwich. Ralph married Eve, and they have a daughter, Diana, who is now married to Ramon Bennett with two sons, Dominic and Julian.

Peter, mother and I continued living at 13 Temple Sheen after my father died. Peter and I both went to Willington School in Putney until we took our Common Entrance exam to St. Paul's School, Hammersmith.[4] Because Mother had very little money, she went to see Willington School's headmistress, a formidable woman named Miss Warren. She had white hair, wore a gown and terrified all staff, pupils and parents. She insisted on teachers being addressed as "Sir" even though the teachers were female, with one exception, because all the male teachers had been called up into the armed forces (we had the one male master, after Dunkirk, because he was invalided out of the army—permanently doubled up having spent 36 hours in the sea waiting to be rescued). Mother's reason for seeing the terrifying Miss Warren was to enquire whether there was any chance of either my brother or I obtaining a scholarship to St. Paul's. Miss Warren replied, "Mrs. Cooper Thompson, you cannot put more into a pint pot than a pint!"

Living in Temple Sheen, we had a few friends who lived nearby who were very kind to our depleted family. The Chestertons lived opposite; James Chesterton was Peter's age and went to Marlborough College, which I visited once to see him, probably around 1950. It seemed a very long way. Helen was his younger sister. They were related to G.K. Chesterton the writer (an Old Pauline[5]).

The Drivers, Bobby and Marjorie, also had a daughter and son, Sally and Nick. Sally was my age, and Nick was younger. He subsequently went to Collet Court, the junior school for St. Paul's, then on to St. Paul's (his father had also been to St. Paul's). Then there was the Wells family, who had an only son, Jonathon, a good friend of mine who also attended Willington. His father sometimes drove us to school in his smart sports car with the roof down. I normally took the bus or later cycled.

When I was about 12, I had a bad bicycle accident and knocked all my front teeth out. In the holidays we would sometimes visit my mother's mother and sisters. We would drive down past Heathrow, which at that time was a field with a few hangars and small planes. I always enjoyed those visits because our relatives lived in a large, old farmhouse in Bracknell. If Peter and I were to stay for a few days, they would meet us at the station in their pony and trap. Bracknell was then a small market town. On Saturdays there would be a cattle market in the center of the town. This was always great fun, seeing the cows, bulls, sheep, chickens and geese. My aunts had two Jersey cows, which we used to milk, and we learned to make butter. They always had many dogs. Aunt Yvonne bred champion bloodhounds and Oona bred pugs and poodles. They also had horses and kept chickens. I loved collecting the eggs. They had two other horse-drawn carriages in addition to the pony cart and two cars. They had a big house with lots of different rooms and a dairy. There was a fox's head in the hall, and, in a second sitting room/morning room, through which we had walk to get to the loo,[6] they had a huge stuffed grizzly bear. It seemed about six feet tall. It was scary going to the loo because the bear watched you as you crept past him. There was also a cuckoo clock on the wall, which made you jump out of your skin. There were lots of bedrooms and enormous baths and bathrooms. In fact, all those bedrooms, many unoccupied, made it very scary going up to bed with the huge grandfather clock ticking and chiming!

I also enjoyed the few occasions when we visited my mother's aunt—Phyllis Horlick. She was my grandmother's younger sister who married Oliver Horlick (chairman of Horlicks[7]). They had a beautiful Queen Anne house in Marlow called Seymour Court. They had a cook and a maid who waited on us at table—very exciting! There was a full-size snooker table just off the library. In the grounds they had a beautiful garden with wonderful herbaceous borders, three greenhouses (one of which had orchids), a grass lawn tennis court, a lake and woods. Needless to say they had two or three gardeners. Aunt Phyllis had a daughter Ann, just younger than mother, and she was married to Colin Gordon and had a son Jeremy, who was just a bit younger than me. I played a lot of tennis with the family. Jeremy and I have remained in touch, and he is married to Liz, with three children.

2. Willington School and St. Paul's School

Mother devoted her life entirely to looking after Peter and me. She had very little money because my father had not expected to die, and she had to manage on an extremely modest income. Meals were very basic—cottage pie, boiled fish and parsley sauce and tripe and onions with cheese sauce! We were all on post-war food rationing. Our house was built before the war with very modest heating by radiators only on the ground floor, except Mother had a gas fire in her bedroom. So in the winter when it became cold, I used to sleep well under the bed clothes, venturing out in the morning with the windows frosted and even the sheets stiff with frost!

Both Peter and I duly sat our Common Entrance examinations for St. Paul's, followed by interviews. To mother's delight and relief we were both accepted. But she told us that, owing to our financial situation, we would have to leave at the end of the summer term when we turned 16.

I thoroughly enjoyed my time at St. Paul's. I was hugely impressed with the great building, and I will never forget the hair-tingling experience of attending assembly in the Great Hall. At 9:00 a.m. the High Master and the Captain of the School (the Head Boy), wearing their gowns, would walk down the two aisles towards the stage in front of the huge organ. Brief prayers were said in Latin followed by a moving hymn where 600 boys and the 30 or so masters would sing at full volume. Many of the hymns and, of course, the Lord's Prayer were also spoken in Latin. But there were a few special memorable hymns like "Onward Christian Soldiers" and "Jerusalem," which were thundered out of the organ with all boys on maximum volume! Fantastic and unforgettable.

I started in Form 4b in 1952. I did well in my first year, particularly in maths and science, so at the end of the year I was moved up to 5x—a scholarship form specializing in science. I had asked to be put into 5x rather than 5a, 5b or 5c because I had already decided that I was going to be an engineer and I needed to push on with science and maths. Although we continued to do Latin, history, English, French etc., I showed little interest in the other subjects.

In sport I was reasonably good at rugby and cricket, although not good enough to make the first team. At St. Paul's we all had to join either the Combined Cadet Force (CCF)[8] or the Scouts.[9] Nearly everyone joined the CCF, which was divided into army, navy or air force, and every Monday we had to go to school dressed in our uniforms. So on Sunday evenings you had to polish your boots, blanco your gaiters and belts and press your shirt, jacket and trousers. It was great fun dressing up as soldiers and traveling on the bus. On one Monday lunchtime I had been shooting in the rifle range when suddenly in came Field Marshall Montgomery, who was an Old Pauline. We all stood to attention. As I had done well, all the shots in and around the bull, I was immodestly holding my target in front of me. Montgomery said, "You're a good shot. Have you had lunch?" "Yes, Sir." "Quite right, never take your troops into battle on an empty stomach!" He regularly visited the school, occasionally on Mondays to inspect the cadets or attend a passing out parade. (The school buildings had become the headquarters of the Home Forces in July 1940, and, in July 1943, the headquarters of the XXI Army Group under the command of General, later Field-Marshal, Bernard Montgomery. So it

The physics laboratory at St. Paul's School, Hammersmith, in 1952 showing Form 4b: Left to right: front row—Smith, Williams, Knott, Matheson, Sumner; second row—Barr, Thompson, Wild, Davy; back row—Twine, Anon, Westerman (author's collection).

was at my old school that the XXI Army part of the military side of the invasion of Europe was planned, including the D-Day landings.)

We played rugby in the autumn, boxing in the spring term and cricket in the summer. In boxing I was one of the best for my weight and consequently boxed for the school on a number of occasions. I particularly remember boxing against Harrow, Wellington, Ardingly and Arborfield. I was awarded my club and school colors for my boxing. We were taught well—to counter box, weave, feint and duck. I won most of my matches in this way.

Below are Bo Langham's notes (he was the boxing coach and physics teacher) of my box against Wellington in 1953. These notes were given to me in 2012 by a fellow Old Pauline who had been examining sporting archives. I had been invited to make a speech as an old boy at a St. Paul's Vintage luncheon, and I am including them here as it is a record of one of my few sporting achievements!

Bo Langham's Notes:

1953 N C Thompson (13.8 yrs–5.12 stone) beat Caldecott (15.1–6.0) Wellington on points.
The weights given above were those interchanged (between the two Schools) prior to the match. At the weigh-in Caldecott was 6.5 and Thompson 5.13.... (Pros and cons for Smith/Thompson to meet C followed.) The next paragraph I quote seems to illustrate the cliché "History repeats itself."
It seemed to me that either of the Paulines was bound to lose badly, but that—all things

2. Willington School and St. Paul's School

considered—Thompson might minimise his punishment better than Smith in this particular match. I therefore asked Thompson to take it on, and he was under no delusion about the nature of this task—I told him he was "for it!" (Some schools try to give an inexperienced boy a false sense of security by telling him that he has "a soft job on," when they know it to be untrue. They feel sure that the boy will put up a better show than if he enters the ring 'scared.' They may or may not be right for their team.) I am sure that a Pauline, who has been taught to (try to) outbox his opponent, however tough, gives of his best—in defeat or victory—if he knows he is the best available boy for the job. What better example could one quote than Thompson in this bout? Any thought of a win must have been the wildest "pipe-dream" in his mind.

At the weigh-in I was badly shaken (doubtless Thompson was—inwardly, though he did not show it). I had a hunch that it was best to leave the pairing (Thompson v Caldecott; Smith v Comerford) as shown on the programme, although an interchange would have made a better paper match. How delighted I was! We won both bouts!

Nigel Thompson aged around 14 (author's collection).

Thompson set out to avoid defeat by accurate timing. Caldecott tried to make a fight of it. The harder he tried the harder he ran on to Thompson's left. Finally he flung caution to the winds and endeavoured to win by thugging it. Thompson, still watching carefully, took every opening and kept Caldecott away. Caldecott, now thoroughly annoyed with himself, and (it's my guess!) really hurt for the first time in the ring, lost efficiency. Thompson, as cool as ever, made it all look quite simple and gained one of the most creditable wins of the team. It was the first straw in the wind.

We won this match by 12 bouts to 4. Wellington were a strong and confident side with an unbroken series of wins to their credit. Comments overheard in the changing room betrayed their confidence in winning this match also. The reasoning was simple. They had beaten Harrow for the first time ever, and we "had only drawn" with them: therefore Wellington ought to win comfortably.

I list below the notes taken by the three judges of the three rounds:

- Caldecott lands some right swings, but Thompson both leads and counters well.
- Caldecott tries to fight, but Thompson keeps him off with lefts. Caldecott is still swinging. Both are tired towards the end.
- Caldecott leads, and Thompson counters. Few obvious blows are struck—but

Caldecott is checked by that accurate counter.
- It is pretty even. Caldecott, more fiery, starts with both hands, but his rights are ineffectual. Thompson counters well with good lefts.
- Caldecott slaps fiercely with his right, lowering his head the while. Playing safe, Thompson steps neatly back, and again Caldecott's rights do little or no damage.
- Caldecott is most persistent, but Thompson dodges very nicely, and again counters well, mainly with his left hand.
- Caldecott leads lefts followed by some swinging rights. Most are stopped on the glove. Caldecott's round.
- Caldecott swings with rights again, and is cautioned for the open glove. Thompson is plucky, and stands up well, but is driven back. An even round.
- A good round. Caldecott is physically the stronger, but Thompson takes most of his swings on the glove, and still scores. Thompson's round.

I have included the above because I was given these notes in 2012 from a fellow Old Pauline who had come across them in an archive he had been researching and they are complimentary about my boxing skills!

My final year at St. Paul's was 1955. I was in 6x and I took my GCE "O" levels and passed General Maths, Additional Advanced Maths, Physics, Chemistry, English and French. I failed Latin and History which was not a surprise as I made no effort whatsoever in those subjects. I had decided that they would be of no use to me!

During the winter months the fog was often so bad in London that the bus conductor had to walk in front of the bus to keep it on the road. This was because all the houses and offices in London burned coal.

3

Starting Work and Rhodesia (1955–1960)

In 1955 my mother was introduced to Dennis Goodban, a bachelor banker just returned to the UK from the U.S. where he had been the Midland Bank[1] representative in New York for three years. Prior to that he had the marvellous bachelor job of being one of the bankers on the passenger ships crossing the Atlantic. He made 98 crossings of the Atlantic on the *Queen Elizabeth*, *The Queen Mary* and the *Mauretania*; travelling first class, of course, all at the expense of the Midland Bank. On returning to England he became a regular attender for Sunday lunches at 33 Gilpin Avenue, East Sheen. (In 1954, owing to a shortage of funds, Mother had had to sell Temple Sheen and downsize to Gilpin Avenue.)

I left school in the summer of 1955—just aged 16—and started work in September as "office boy" for Robert D. Ward, MI Struct E, Structural Engineering Consultants, 7 Ely Place, London. There were three qualified structural engineers in the office and me. We worked six days a week 9:00 a.m.–5:30 p.m. except Saturdays when we stopped at 12:30 (half day). My role was to get in early—open up the office, light the gas fires and put the kettle on. I had my own drawing board, and I quickly learned to draw in ink with ruling pens and became a competent draughtsman. I had to take the drawings for printing, the letters for typing and the completed prints to the architects and contractors. I attended night school four nights a week—learning theory of structures, "A" level mathematics, strength of materials, geology, electricity, hydraulics, structural steelwork and reinforced concrete design, building construction and materials. I worked very hard. I was paid £3.00 (US$8.40 at 1955 exchange rates) per week, but at Christmas I was given a bonus of £5.00 and, because I had worked so well and they were pleased with my progress, my salary was increased to £4.00 per week. During the first six months of 1956 I also learned to do simple steelwork and reinforced concrete design and detailing. We used slide rules for all our calculations.

My brother Peter also left St. Paul's School when he was just 17 in 1954, and he joined Lloyd's of London[2]—the big insurance brokers—as a junior. He also made good steady progress.

Our first jobs were both down to our Mother's efforts. Mother's father had been a major Lloyd's underwriter and, although my grandfather had died before I was born, Mother approached his old firm and persuaded them to take Peter on as a junior. For me, as I wanted to be either an architect or an engineer, Mother became friendly with a couple living in East Sheen—the husband was a structural engineer and drove a new Rolls Royce! He also owned a steelwork fabrication shop. It was his company, Robert D. Ward, that I joined as "office boy!"

Since our father died in 1948, Mother had spent her life entirely focused on my brother and me, on our schooling and then getting us both started in careers. We owed her a lot. Early in 1956 Mother married Dennis Goodban. Both my brother and I quite liked him, but we realized fairly quickly that it was really for his career and for his public image that he wanted to be married with a family. I discovered before my brother or mother that Dennis was probably more interested in my brother and me than in our mother. Mother did not realize that Dennis was "gay" for quite some time. Of course, this was the period when to be "homosexual" was against the law and an imprisonable offence. (Homosexual acts were not decriminalized in England and Wales until 1967.)

Mother and Dennis appeared to be fairly happy. But much later, during questioning from me on "how was married life," Mother said she was not unhappy but the marriage would never be complete as he was not like most men, and they were not able to have a physical relationship. I knew what Mother meant, but I also knew that he was one of those that at school we would call "homos." On one occasion, after playing a game of squash, Dennis and I were in the changing room having a shower, and he suddenly appeared close to me asking to borrow the soap and exposing an erection. I turned on the cold water and disappeared out of the shower at speed! Nothing was said about this embarrassing incident, and I did not tell my mother or anyone else.

In the spring of 1956 Dennis Goodban, our stepfather, was appointed to a senior position in the Bank of the Federation of Rhodesia and Nyasaland with the head office in Salisbury, Southern Rhodesia. I am sure that being a respectable married man helped him land this position! Prior to this he had worked for the Midland Bank for many years, and he had very much enjoyed his most recent position as the bank's representative crossing the Atlantic.

So Dennis spoke to both my brother and me saying, "I am taking your mother out to Rhodesia, and if either or both of you would like to come with us, here is a one-way, first-class ticket on the *Durban Castle* due to sail from Tilbury Docks in London in June 1956." He added, "You will have to buy yourself a dinner jacket as they dress for dinner every evening, and the journey will take three weeks to Cape Town." Not surprisingly we both jumped at the offer—I had always wanted to go to Africa. I duly bought my first dinner jacket—£10 from Burtons!

The *Durban Castle* was an intermediate Union-Castle mail ship.[3] We stopped at Rotterdam, Las Palmas in the Canaries, then Ascension, St. Helena and Cape Town. Aged 16, this was the first time I had been abroad. I had my seventeenth birthday on board ship shortly after crossing the equator. Peter and I shared a cabin. Being first class we had our own "en-suite"—a saltwater shower. There was no air-conditioning in those days.

We found the journey out by sea enormous fun—rough seas, huge menus, wonderful food, dressing for dinner, gin and tonics in the bar, playing bridge after

3. Starting Work and Rhodesia

dinner. There were four Cambridge botany undergraduates on board who were forever catching fish and dissecting them. We enjoyed the sun, playing deck tennis[4] and swimming. I nearly won the deck tennis competition—just beaten in the final!

We went ashore at all the ports where we stopped except Ascension where only one passenger, who worked for Cable and Wireless, went ashore. He was lowered over the side in a basket onto a tender waiting for him. Thousands of evil-looking "devil fish" surrounded the ship devouring anything that was thrown overboard in seconds—in such a fury that the sea appeared to boil!

The next stop, after about two weeks at sea, was the island of St. Helena. An exciting experience. We explored St. Helena on an amazing old bus. Where Napoleon was exiled and died, we saw forts still with their original cannons to repel any French fleet trying to rescue their emperor. We went through four different climate zones. I loved the island. In the harbor there were donkeys laden with flax to load onto our ship to take back to England to turn into hairy string for the British post office. During the voyage we had enjoyed watching the flying fish breaking from the bow waves and flying with dazzling colors—20 to 50 to 70 yards before again entering into the ocean. I was not to revisit St. Helena for the next 50 years.

The road approaching Birchenough Bridge over the Sabi River near Chipinge, Manicaland, with a young baobab tree visible on the right. The arch bridge, just visible in the distance at the center of the photograph, is named after the chairman of the Beit Trust which funded it. The bridge was opened in 1935 and is two-thirds the size of that across Sydney Harbour, also by designer Ralph Freeman (1974, photograph by Colin Weyer, www.rhodesia.me.uk).

We arrived in Cape Town at the end of June 1956. My stepfather had brought a new car with us on board ship. On stepping ashore we climbed into the car and started a four-day drive up to Bulawayo to stay with some friends before driving on to Salisbury to commence our new life in Southern Rhodesia. Having driven through intense rain and hail through the mountains, we entered the Karoo where we stopped for our first night in Laingsburg—the Grand Hotel! Laingsburg was a very small town with only one road going through it. Outside the hotel was a railing to which a horse was tethered—it was exactly like the set of a classic cowboy film. Driving through the Karoo was amazing—the road just went on and on and on. Our second night was spent at Bloemfontein where we had ice on the car in the morning, but by midday the sun was warm. For the third night we stopped just over the border into Southern Rhodesia where we were beginning to see some game—mostly the occasional buck or a warthog family dashing across the "strip" roads. They were so named in Rhodesia because the road surfacing was two strips of tarmac each about 60 cm (approximately 2 ft.) wide. When you met a car coming towards you, each driver had to move to the left onto only one strip—an exciting new experience.

An old strip road south of Fort Victoria (now Masvingo) close by the ruins of Great Zimbabwe. The strip road, with two narrow, parallel strips of asphalt, one for each wheel, was a uniquely Rhodesian concept used from 1933 as a low-cost means of opening up the country for development—1,174 miles (1,890 km) built by 1938 and more than 2,000 miles (more than 3,300 km) by 1945. Most strip roads have now been replaced by full-width asphalt roads, but some still remain in remote areas (1974, photograph by Colin Weyer, www.rhodesia.me.uk).

3. Starting Work and Rhodesia

In Bulawayo we spent two nights with a family whose mother had lived in Temple Sheen throughout the war. They had a welcoming dinner party for us, and this was the first time I was to meet a senior political figure, Garfield Todd, who, at the time, was Prime Minister of Southern Rhodesia. That evening I also met Jack Haviland, a civil engineer, who had just opened a structural engineering office in Salisbury with Reginald Bray. I was also given the name of another structural steelwork practice in Salisbury for me to approach. A very fortunate evening for me! In the newly formed Federation, construction was booming so unsurprisingly, the following week, when we reached Salisbury, I was offered a job by both practices as a junior designer detailer at £5.00 per week.

At that time very good cigarettes which were made with local tobacco were priced at 2s. 3d. (about U.S. 43 cents at 1956 exchange rates) for 50. Everyone smoked and carried a packet of 50 with them in their shirt pocket as today everyone carries a mobile phone!

So, I started work immediately for Haviland Marshall & Bray, where I worked for nearly four years until I returned to England on a Federation of British Industries (FBI)/British Council Scholarship. There were only two of us in the office—we both had two drawing boards because we had so much work. Initially Reg did most of the calculations and design work (all on a slide rule) while I did the layout drawings on one board with the reinforcement or steelwork details on the other.

I attended night school at Salisbury Polytechnic four nights a week where the chief municipal engineer, a Scotsman, ran a class for young engineers to take their professional qualifications: AMIStructE and AMICE (Associate Member of the Institutions of Structural and Civil Engineers) There were at most six of us, all at different stages. We would sit there for two hours copying out his notes and trying to understand how to do the calculations. I would have to spend one day most weekends studying at home trying to understand the week's notes.

Life was good. I learned to play golf—starting with a handicap of 16. I played at Ruwa Country Club where I also played a lot of tennis and a little cricket.

Dennis's sister, Ann Seymour Smith, with her husband had been early settlers and in the 1930s had built a fantastic hotel high up in the Bvumba mountains near Umtali, which they named Leopard Rock. We enjoyed staying there many times, playing golf and riding horseback. We just had to watch out for puff adders in the bunkers! It was at Leopard Rock where I first met Sir Edward Whitehead who later became Prime Minister of Southern Rhodesia as the federation was beginning to fall apart and just before Ian Smith took over, declared UDI and removed Rhodesia from the Commonwealth. Edward Whitehead owned a farm in the Vumba, where he lived with his cousin Matt in an old farmhouse. When the Queen Mother visited the federation, she came to stay at Leopard Rock Hotel specifically to see her friend Sir Edward. He was so embarrassed about the internal state of their house that they took the roof off, so they had to entertain the Queen Mum in their garden.

We used to play snooker in the bar with Edward after dinner, where he introduced us to his favorite liqueur "Green Chartreuse!" ("Green Chartreuse, 110 proof or 55 percent ABV, is a naturally green liqueur made from 130 herbs and

The Leopard Rock Hotel, built in the late 1940s in the Vumba Mountains, Eastern Highlands, nestles at the foot of the rock after which it is named. Queen Elizabeth, the Queen Mother, of Britain said of this place, "There is nowhere more beautiful in Africa" (1976, photograph by Colin Weyer, www.rhodesia.me.uk).

Nigel Thompson, as a teenager working with Haviland Marshall & Bray, designed and supervised his first "tall building," Fanum House (completed 1959), which is the AA head office, in front of the Pearl Assurance Building (pearl on top of tower) to the right of Jameson Avenue (now Samora Michel Avenue) in Harare (1977, Colin Weyer).

3. Starting Work and Rhodesia 23

other plants macerated in alcohol and steeped for about eight hours. A last maceration of plants gives its color to the liqueur" [Wikipedia].)

I gained a lot of experience in the building of Salisbury's first tall buildings both in design, working with architects and supervising the construction on site, which included checking the depth of piles and that all the steel reinforcement had been fixed in the correct position before concreting. I helped build several tall bank buildings, two cinemas, a dairy, university buildings etc. I was also lucky enough with other local engineers to visit the Kariba Dam[5] which was under construction by the Italian contractors, Impreglio.[6] We were able to go right down into the cofferdam deep in the Zambezi to inspect the amazing dam construction. It was incredibly impressive. We had flown up from Salisbury to Kariba in a small Dakota plane which had difficulty in landing on the first attempt because there were two lions lying on the grass runway!

Some weekends, with my friends, we would drive through the night in a

The Kariba Dam is one of the world's most impressive man-made structures. The 420 ft (128 m) tall, double curvature concrete arch dam in the Kariba Gorge of the Zambezi River basin, was built to generate hydroelectricity for Zambia and Zimbabwe. Arch dams are designed so that the water pressing against them compresses and strengthens the structure as it pushes into its foundations. Such dams are often chosen for canyon or gorge locations, as in this case, because steep walls of stable rock help support the dam structure. The 1,900 ft (579 m)-long Kariba Dam forms Lake Kariba, which extends for 170 miles (280 km) and holds 150,000,000 acre-ft (185 km^3) of water (1975, Colin Weyer).

VW Beetle to Beira where we would sleep on the beach, sunbathe and surf in the Indian Ocean. I would spend weekends with friends on their family tobacco farm where we would go shooting game. We also spent time in Mozambique in the very wild Gorongosa Game Reserve—amazing experience living with the wild and dangerous game.

In 1959 I was called up to do National Service in the Royal Rhodesia Regiment subsidiary of the King's African Rifles. We were based at Heany Barracks near Bulawayo. In the army we first learned how to fire and clean our rifles, to carry out simple manoeuvres in the bush, to march up and down the parade ground and endlessly clean and press our kit and uniforms. There were 18 of us in our barracks. To my surprise six admitted to having either been in prison or in reform school. We were all from totally different backgrounds, which was a great experience, but, sadly and unsurprisingly, at that time we were all white! There were also black African units at Heany, and some of the enterprising black soldiers would come to our barracks in the evening to offer their services for pressing our uniforms brilliantly.

After only three weeks we were woken up at 3:00 a.m., issued with rifles, live rounds and snake bite serum. We were taken into the bush to guard a prison. The federal government under Sir Roy Welensky had declared an emergency, and the Army had arrested Dr. Hastings Banda from Nyasaland, Kenneth Kaunda from Northern Rhodesia and Joshua Nkomo from Southern Rhodesia. They were all brought to this prison, and we were to provide a show of strength by guarding the outside of the prison walls. We were lit up, highlighted against the walls as we walked up and down with a loaded rifle in our hands, nervously peering into the darkness. We were only on guard duty for a few days until they were happy the prison was not about to be stormed!

After my military adventures I returned to work in Salisbury and passed my graduateship examinations, and, upon advice from my

Mr. Nigel C. Thompson, the young Salisbury engineer recently awarded a scholarship by the Federation of British Industries for two years' practical training in London, sails for Britain today.

His scholarship covers passages to and from the United Kingdom and his maintenance overseas.

He is shown here (left) saying goodbye to the United Kingdom High Commissioner, Mr. M. R. Metcalf, in Salisbury.

Mr. Nigel Thompson, aged 20, being awarded an FBI (Federation of British Industries) scholarship by United Kingdom High Commissioner Mr. M.R. Metcalf in Salisbury, Rhodesia, 1960.

3. Starting Work and Rhodesia

evening class lecturer, I applied for Rhodesian citizenship. I won a prize from Salisbury Polytechnic for being the best student of the year. I then immediately applied for the FBI (now CBI) British Council overseas scholarship for two years' further training at a top British firm in England. I won the scholarship and was placed with Ove Arup & Partners. In January 1960 I was invited to the United Kingdom High Commission where the high commissioner presented me with the award. It was the first time I had had to make a public speech which was reported in the press!

4

Return to England—Scholarship and Joining Arup (1960)

Because I was now Rhodesian, there were educational opportunities open to me that weren't available to young people in England. I applied for and was awarded an overseas scholarship with the Federation of British Industries (FBI), for training with Ove Arup & Partners. I was among the last FBI scholars because that acronym had been usurped by an agency in the U.S. Even though the Federation of British Industries was the older organization, it was forced to change its name, becoming the Confederation of British Industries (CBI). So, from being among the last FBI scholars, I was now among the first CBI scholars.

One day, while at work in Salisbury, I was amazed to get a phone call from Peter Dunican,[1] the senior managing partner of Ove Arup & Partners. Peter was in Southern Africa on business. He talked with me about the firm and said that he would look forward to seeing me in London.

In February 1960 I boarded the *Lloyd Triestino* passenger ship in Beira to sail up the east coast of Africa to Venice on my return to Britain. I enjoyed the sea trip enormously, stopping and going ashore at Dar es Salaam, Tanganyika, where we walked through the tea plantations, then Zanzibar with the strong smell of cloves and spices. The harbor was filled with dhows. The place and the island were magical and amazing. Next we stopped in Kenya, at Mombasa, where some of us went up to swim off Nyali beach. There were no hotels, just the most wonderful soft, almost-white sand and the beautiful, warm sea. Heaven! After Mombasa we stopped briefly at Mogadishu in Somalia and then up to the Gulf of Aden. We went ashore at Aden, another fascinating port with a recent history of unrest, and then we sailed up the Red Sea until we reached the Suez Canal. Some of us got off so that we could take a taxi up to Cairo to visit the world-famous museum and take a camel ride up to and around Cheops Pyramid and the Sphinx.

We rejoined our ship at Port Said, fighting our way through crowds of young men trying to sell us stiletto sword sticks hidden inside a walking stick, other trinkets or their sister! It was now our final leg—crossing the Mediterranean first to Brindisi at the base of Italy and then up the Adriatic and finally up the Grand Canal in Venice. What a fantastic end to an exciting voyage, to land in one of the world's most beautiful cities. After one and a half days of exploration I took a train to Paris where again I enjoyed and absorbed the excitement of that city. So

4. Return to England—Scholarship and Joining Arup

1960: FBI scholars: back row from left to right—(3) Alberto Jesus Rivera from Mexico (6) Nigel Thompson from Rhodesia (7) Fernando Garnica from Bolivia; Fifth row from left to right—(8) Salman Hussein Abu-Sitta, a Palestinian from Jordan (F.B.I. Overseas Scholarships Eighth Annual Reunion, March–April 1962).

in March 1960 I arrived in London to begin my FBI/British Council Scholarship with Ove Arup & Partners.

First I rented a room with a family we knew in East Sheen. Then I arranged to meet Peter Dunican, the senior managing partner at Arup. He told me to come to 13 Fitzroy Street on Friday afternoon where he was joined by Frank Coffin and Bob Hobbs, who suggested that they should put me on to the sixty-strong Sydney Opera House team. I said I preferred the idea of a smaller group as I was used to working in an office of two! So Peter put me on the team for the new National Sports Centre at Crystal Palace!

Arup was not then among the biggest consulting engineering practices, but it was growing rapidly. Compared with what I was used to, it was huge. So, on my arrival in Arup's Fitzrovia offices, I felt completely overwhelmed. At that time Ove Arup & Partners employed a total of around 300 staff worldwide.

I started with Poul Beckmann's group working on the new stadium at Crystal Palace. Poul was a brilliant engineer, and he acted as my mentor. His group was on one half of the fifth floor of No.13 Fitzroy Street. On the floor there were three long rows of drawing boards, each with a desk beside it. Despite the size of the boards there were twice as many engineers on each floor as there would be in the late '70s. Adjacent to the lifts there were two offices, one for Poul Beckmann and the other for Vic Kemp, and two secretaries (one secretary for Poul's group and the other for Vic's). We would prepare all our layout drawings in ink on

Sydney Opera House: Architect—Jorn Utzon; Engineer—Ove Arup (photo Harry Sowden).

4. Return to England—Scholarship and Joining Arup

either linen or drawing paper. The pens we used were either ruling pens or graphos, an improvement on the ruling pen, and we drew with the aid of wooden T-squares, triangular set-squares and protractors. For reinforced concrete construction we would draw elevations of all beams and columns and plans for slabs upon which we drew the reinforcement clearly showing numbers, diameters, bending and positioning of all reinforcement. We would then sit at our desks and prepare bending schedules (it was not until the end of the '60s that Arup, for efficiency, set up a central reinforcement detailing group). For structural steelwork

The National Recreation Centre, Crystal Palace, London—Nigel Thompson, as a young Arup structural engineer, designed and detailed the unique and very popular high diving board for the pool hall (Architectural Press Archive / RIBA Collections).

design we would again prepare all layout drawings showing size and location of all steelwork sections supported by detailed drawings of all connections, fixings, bolts, rivets and welding requirements. It was a different world in the '60s, and we prepared all our calculations with the aid of slide-rules. After about a year in Poul Beckmann's group I had the opportunity of working on my own small project for Vic Kemp, also under Peter Dunican, which I enjoyed very much, as Peter was always interested in what I was doing. He took me to meetings with him and demonstrated the Arup philosophy of collaborating with architects and giving your engineers responsibility. I then started designing precast concrete housing for the chief architect of London County Council. At the time we felt we were helping solve the housing problem and "building the New Jerusalem." In fact we were tearing down dilapidated terraced Edwardian houses and replacing them with medium rise (12 stories) precast concrete monsters! It was a mistake that planners were making in the '60s. We should have restored the terraced housing and thereby avoided breaking up and destroying the communities.

Peter Dunican was clearly pleased with my performance as he appointed me to be one of the Arup tutors for the young architects studying at the Architectural Association in Bedford Square, advising people such as Richard Rogers and

From left: Peter Dunican, Ove Arup—two of the biggest influences on my life (Arup).

4. Return to England—Scholarship and Joining Arup 31

Robin Sutcliffe on structures. I kept in touch with many of the students, and some I continued to work with. Being on the fifth floor of 13 Fitzroy Street, we were on the floor immediately below the Arup Partners so we would frequently meet them in the lift or the loos. Whenever you met Ove he would say, "Who are you?" and "What are you working on?" By the mid–1960s I was a fully qualified associate member of the Institution of Structural Engineers and of the Institution of Civil Engineers. I was appointed a senior engineer at Arup and led a team for the design and construction of Northwick Park Hospital,[2] a brand new 600-bed district general hospital combined with a 200-bed clinical research center. This was another of Peter Dunican's projects, and I worked closely with John Weeks,[3] partner of Llewelyn-Davies Weeks.[4] This was the first of my major hospital projects and was, at that time, the biggest building project in Arup in London.

1963: Nigel Thompson working at office desk, with elevations of Northwick Park Hospital on wall (author's collection).

5

Friends and London in the Early 1960s

In the early 1960s I had found a flat in Belsize Park and invited John Havergal to share with me. We paid a total of £2.50 a week for the first floor of a Victorian house on Haverstock Hill. Initially I was still studying four nights a week at Westminster Technical College, doing my higher national Certificate for my civils membership. But on Saturdays with friends we enjoyed '60s London.

I had first met John Havergal in Southern Rhodesia. He had arrived in the Federation of Rhodesia and Nyasaland in 1958–9 after hitchhiking his way from Woolton Hill in Hampshire across Europe, then Tunisia, the Sahara, Somalia, Kenya and finally Rhodesia. We had become friends, and I called him when I arrived in London in 1960. On a couple of occasions John and I went climbing mountains in the Alps—The Dents du Midi near Champery. On one occasion, when we were staying in the top Mountain Lodge at Susanfe, we met Dux and Monique Schneider[1], who became good friends. Dux and Monique were a very interesting couple.

Monique, a writer, was the daughter of two Jewish psychoanalysts working with Sigmund Freud who had fled to America from Hitler. Dux, most unusually, had been educated like me, first at Willington School in Putney then St. Paul's School in Hammersmith. After school Dux went to the United States where he joined the U.S. Army near the end of the war and had the horrific experience of being one of the U.S. soldiers who opened the German Belsen Concentration Camp. Although Dux and Monique spoke English, German and French fluently, Dux could never bring himself to forgive the Germans.

I used to spend some weekends in the country in Hampshire and Berkshire with John Havergal and his family. On other occasions I went rock climbing in Derbyshire with Dux and Monique or to concerts with Mike and Ann Hill, whom I had also known in Rhodesia. Through the FBI Scholarship I developed a close friendship with Salman Abu-Sitta, a Palestinian officially from Kuwait. From Salman I learned an enormous amount about the problems arising from the Balfour Agreement and the appalling treatment of the Palestinians by the Israelis, who drove families off their land and did everything they could to destroy Palestine and the Palestinians. It is a very sad irony that the Israelis should behave in this way after suffering themselves so horrendously under the Holocaust.

5. Friends and London in the Early 1960s

The Confederation of British Industries Scholarship Office arranged a few meetings for scholars to meet key members of British industry. On one occasion they invited Peter Dunican from Ove Arup & Partners to join a panel of industrialists and professionals to tell us about what they were doing followed by a question and answer session. So obviously I asked Peter Dunican to expand on the role of multidisciplinary engineering design. His favorite subject! A friend said to me afterwards, "You are now destined for promotion!" On another occasion, which was probably the highlight, His Royal Highness the Duke of Edinburgh came and told us about his time in the Royal Navy and his experiences in visiting the countries from where the CBI Scholars had come. He then talked to each of us, asking about our ambitions.

I had two other FBI scholarship friends, Alberto Jesus Rivera from Mexico and Fernando Garnica from Bolivia. When John Havergal moved to the north of England for his frozen food business, I invited Alberto to join me at the flat. He persuaded me to move to better premises further up the hill into Hampstead Village. He was great fun to share with, and he taught me to cook Mexican pancakes for breakfast.... "Maria ... desayuno!" He had a delightful Swedish girlfriend, whom he later married. As both of them had limited English, they communicated through their own version of our language. After about 18 months it was Alberto's time to return to Mexico, so Fernando became my flatmate. Fernando loved the English countryside and enjoyed what he described as typical British behavior. He told a story of how he loved the British character. When he visited a pub in the countryside there were several old men sitting drinking their pints of beer. No one was talking to anyone until a man came in with a dog, and then everyone talked to the dog!

In the early part of summer 1963 I went with Sam Price, a colleague from Arup also working in Poul Beckman's group, and a friend of his, Philippa, on a tour around France in Sam's Bentley. We visited Chartres, then many of the chateaux of the Loire Valley. We enjoyed driving the old Bentley all around France and, particularly, along the Grande Corniche on the Riviera. Of course, on our cultural tour, we naturally went to see two of Le Corbusier's great projects, La Tourette[2] and Ronchamp.[3]

6

Meeting Nicky (1963)

On one occasion when I was staying with John at his parent's house, Tower House, in Woolton Hill near Newbury, we were invited to a party at Burlyns by Philippa Bonnett. That evening I had a bit too much to drink so that when I asked to go to the loo, I was directed down the hall, turn left and go straight ahead, which I did, stepping over a piece of wood that crossed my path. There was an almighty crash as I put my knee through a plate glass door! While I was being fussed over by Angela Darling and a few others, I noticed a young girl in her pyjamas laughing! John was keen on Philippa Bonnett and married her in 1963. I was the best man.

South elevation of Burlyn's, East Woodhay, Newbury, Berkshire, the family home of Major and Mrs. E.H. Bonnett—Nicky's parents (author's collection).

6. Meeting Nicky

Shortly after John's wedding I was invited to stay at Burlyns for Julie Havergal's wedding, and it was there that I met an amazing, lively, bright and incredibly attractive young woman—Nicola Bonnett (the laughing girl last seen in her pyjamas when I broke the plate glass door). At that time I had a current girl friend, a Danish girl with whom I had spent the previous Christmas in Copenhagen. On returning from Christmas in the snow, my flatmate Fernando announced his time at Arup was up and he had to return to Bolivia. As he was leaving I said, "You are my guest. What would you like to do?" Fernando replied that he would like to go to the most expensive restaurant in London. We therefore set out by taxi to the Mirabelle, which was, at that time, frequented by people like Christine Keeler.[1] We had oysters, turtle soup, lobster, then steak, crepe suzette and, to finish, coffee and cigars. For wine we had champagne, then white wine, then red and finally cognac with our cigars. The bill was enormous. The total came to £22.00 (US$62 at 1963 exchange rates)! At the time it was a huge bill!

Nicola Bonnett was far too young of course, and I couldn't possibly consider her as a girlfriend! However I now had a reason to spend more time and weekends in the English countryside! I had many wonderful weekends staying at Burlyns going riding with Nicky over the Berkshire and Hampshire Downs, helping her father clear brambles in readiness for shooting, going to the pub and participating in hugely enjoyable shooting parties. In all of these events Nicky was very actively involved. Nicky had moved to Oxford where she was doing a secretarial course and started singing lessons as she had a very remarkable mezzo-soprano voice. I would sometimes drive in my Mini[2] down to Oxford for the evening, and we would visit The Trout Inn,[3] a historic public house in Lower Wolvercote, north of Oxford, close to Godstow Bridge and directly by the River Thames.

Nigel riding out with Nicky (not shown) (author's collection).

7

Summer in Turkey (1964)

In the summer of 1964 Dux and Monique invited me to join them in climbing some mountains in Eastern Turkey. They had decided to give up living in London, buy a Land Rover to explore, climb mountains, live and write books in Turkey. I joined them driving out through Europe. Nicky had wanted to join us, but Dux felt she was too young. He did not want to take the risk or responsibility!

When we were in the eastern part of Turkey I climbed Suphan Dag, an extinct volcano, which was 4,434 m (14,547 ft.), second only to Ararat in height. We had all been due to climb the mountain, but, as Dux fell ill with a sudden fever, "the ascent of Suphan Dag" was left to me. To get into this part of Turkey, around Lake Van and the Hakari mountains, we required a license and had to have an armed Turkish guard with us at all times because it was a Turkish military zone. So the climbing party was four of us: the armed policeman, two local Kurdish guides and me. It was not a spectacular climb, being mostly hard slog over loose shale. The highlight for me was on the way down when the younger Kurd collected his beautiful Arab horse and asked me to go with him, me in the saddle, to where his whole family were camped. We slipped away from the policeman, leaving him shooting at stones and anything that moved.

As we disappeared over the hills, we came upon a cluster of perhaps 25 tents. I was welcomed into the gathering, with the women wailing, and then ushered into the headman's tent. We sat on wonderful carpets, shook hands and nodded and smiled to each other while my young Kurdish guide talked, presumably about the ascent of Saphan Dagh and the mad Englishman! The headman offered me warm sheep's milk and naan bread, a new experience for me, while the whole clan squeezed into the tent to stare at me. After about half an hour of smiling, eating and drinking, I persuaded my guide to find our way back to the edge of a small village where I joined my colleagues. I think the village was called Gevash. From there at Lake Van we visited and greatly admired the small Armenian island church of Akdamar. The tenth-century church is a small cruciform structure of red stone, all of it decorated with friezes and reliefs. Most striking are the Old Testament scenes of an awkward David about to slay Goliath.

After six weeks of extended holiday driving through western Europe, the Balkans and Turkey I had to fly back to England from Trabzon to get back to work. Something I always remember about Turkey, at that time, was that when we heard a funny noise from the engine of the Land Rover and stopped to look

7. Summer in Turkey

Old Testament scene of David and Goliath on the frieze of the Armenian Church of Akhtamar, Lake Van, Turkey (Lostinafrica, 2007, English Wikipedia).

under the bonnet to peer at the engine, even in the remotest part of the country, we would be joined by one or two locals who would have appeared from nowhere saying "Kibris, boom, boom." This, of course, was when Turkey was bombing part of Cyprus.[1]

It was mostly an enjoyable and interesting trip, but being in a Land Rover for six weeks with a couple who occasionally reminded me of *Who's Afraid of Virginia Woolf?* made it a memorable holiday!

8

Senior Engineer and Marrying Nicky (1965)

In 1962 my scholarship training was officially completed, and I should have returned to the Federation of Rhodesia and Nyasaland. But the federation had collapsed. Ian Smith had been appointed Prime Minister of Rhodesia. Northern Rhodesia had become Zambia under President Kaunda, and Nyasaland had become Malawi under Dr. Banda. Smith then announced the Unilateral Declaration of Independence from the United Kingdom. I had nowhere to go back to, and Arup wanted me to stay on. Mother and Dennis had moved to Malawi, but Peter chose to stay on in independent Rhodesia. So as a young engineer I negotiated what I and my colleagues thought was a very good salary—£1,000 (US$2,800 at 1962 exchange rates) per annum. Senior engineers at that time at Arup earned £2,000. I had asked for £1,100, but Peter Dunican said, "I have to pay you enough to keep you happy but not so much that it completely upsets the system!" Friends thought it was a huge salary!

By 1965, as I wrote earlier, I was appointed senior engineer when I had taken on the responsibility for the design of Northwick Park Hospital. The hospital was to be a combined 600-bed district general hospital and a research hospital with 200 laboratories as single bed wards supported by more laboratories and an animal house. John Weeks, the architect, was very influenced by the principle used by Florence Nightingale and Isambard Kingdom Brunel in their design for their hospital at Renkoi in the Dardanelles during the Crimean War. The successful, simple principle was to have a main street to which you clip extendable buildings on either side. The only constant in the design for hospitals is to design for change. Today's research laboratory can become tomorrow's department, so to be able to clip buildings on to a serviced street allows for maximum adaptability, extendability and change. I very much enjoyed working with John Weeks. The project was designed in imperial (feet and inches) with a standard grid throughout. The grid was based on the ideal width of a laboratory and a single bed ward of 11 ft. 4 in. The width of a four-bed ward was 22 ft. 8 in. We then planned the whole hospital on a 2 ft. 10 in. grid being 4 units for the lab. and 8 units for the large wards. The horizontal structure was coffered flat slabs on a 2 ft. 10 in. grid for the coffers and 22 ft. 8 in. for the column spacing. John was keen that the load-bearing façade should reflect cybernetic serendipity, which I managed to achieve

8. Senior Engineer and Marrying Nicky

South elevation of the main ward blocks of Northwick Park Hospital. The two-story building on stilts, on the right hand side, is the operating theater block. The elevation demonstrates the external loadbearing mullions reducing in number as you go higher (Arup).

by placing the mullions on the grid but spaced out to reflect the load they were carrying.

My team were still working on the fifth floor of No. 13 where I shared a secretary with Vic Kemp, Julia Callaghan. She was the daughter of Jim Callaghan, the Chancellor of the Exchequer, who later became Prime Minister. In the '60s Arup held Christmas parties for all staff. Before the party in 1965 Julia invited Nicky and me to come to No. 11 Downing Street for drinks before all going on together to the Arup party. For some security reason we had to enter through the front door of No. 10 Downing Street. That was to be my first visit to that famous building!

One day David Thomas, the LDW architect responsible for designing the ward blocks at Northwick Park asked me to join him and his partner Peter Foggo, who was working for Arup Associates, to help them with some evening and weekend work as their structural engineer. The projects were all housing either designed in structural steel similar to the famous Farnsworth House by Mies Van der Rohe or in precast concrete. At that time I was also asked by Peter Stone, another LDW architect working on the operating theaters at Northwick Park, to join him and his architect partner Royston Summers for some more weekend and evenings private work as their structural engineer. We completed a couple of projects. The first being an automobile maintenance factory in Kings Lynn and the second an award-winning terrace of six, three-story, brick houses on the

Green at Blackheath, London for six friends: Royston Summers, Michael Frayn, the author, and four others.

Having been made a senior engineer at Arup and having a serious girlfriend, I decided that I should not go on accepting private work.

On April 23, 1965, Nicky and I became engaged. Previously I had had many girlfriends, but I was entranced by this amazing girl, young but such a character. I adored her. We had such fun together going to parties and hunt balls, etc. I had never before considered marriage or even living permanently with a girlfriend.

After we became engaged, Nicky's mother, Bridget, told me about Nicky having had cancer of the ovary when she was 13. Initially the surgeons had wanted to remove both ovaries, but Nicky's mother insisted, with the support of one surgeon, that they keep the apparently unaffected ovary. This was agreed, but the surgeon injected the ovary with a heavy dose of radioactive gold. This meant that although Nicky would be able to mature as a woman, it was extremely unlikely that she would ever be able to have any children. Bridget arranged for Nicky and me to visit the surgeon in Great Ormond Street just around the corner from the Arup office. A really nice man, he explained to us what he had done and that, hopefully, she would not have a reoccurrence of the cancer. But, sadly, it was highly unlikely that she would ever be able to have any children.

That didn't bother either of us as neither of us had even thought about children. We were planning to get married in September. My brother Peter got married, I think in May, and then Mother came over to England to meet Nicky and stay for our wedding. Nicky and I met her at the airport, whereupon Nicky had a migraine! After a few days Mother said to me, "Are you sure about marrying Nicky? She is very young, may be ill all the time and can never have any children. You are only 26, and there are many more fish in the sea!"

September 4, 1965: Nicky's and my wedding reception (author's collection).

8. Senior Engineer and Marrying Nicky

Anyway on September 4, 1965, we had an amazing wedding at West Woodhay Church and afterwards at Burlyns. Sam Price, with whom I had recently toured through France in his Bentley, and who had joined Arup a year after me, was my best man.

After the wedding reception, which was just champagne and nibbles, no cake, no speeches, all at our request, we set off in my newly acquired old Land Rover as I had unfortunately written off my Mini six weeks earlier! We had a fantastic meal at Bucklers Hard[1] in the New Forest and stayed at a local hotel. The following morning we had planned to catch a small plane, with our car, from an airport near Lymington. However, my "new" secondhand Land Rover had a puncture on the way to the airport, so, in my smart going-away clothes, I had to change the wheel.

I had originally planned to take Nicky to the Greek Islands, but, owing to a minor financial crisis, I changed plans to drive down through France either camping or staying at "egg-cup" hotels in the Michelin Guide. On our first night in France, staying in the Loire Valley, the landlady was very disapproving because clearly she did not believe that we were married or that Nicky was old enough! As we drove, enjoying the sun and countryside, I loved the old farmhouses while Nicky was admiring all the French horse-drawn carts and rural France. Our target was a small town on the Mediterranean just across the border into Spain. An

September 4, 1965: Nicky and her five bridesmaids in the grounds of Burlyn's; (from left) Diana Dillon, Scilla Atkins, Edwina Bonnett (Nicky's sister); (front, on right) Sally Corke (author's collection).

architect friend of mine had recommended a hotel in the quiet, unspoiled bay where the fishermen still laid their nets out on the beach. When we arrived, having enjoyed France enormously, we were horrified to find signs saying "English Tea," "Fish and Chips" etc. So after one night we decided to escape the horrors of British influence and to explore the Pyrenees, travelling diagonally across the mountains mostly on unpaved roads and staying at farms, until eventually we arrived at Biarritz for champagne and oysters on the Atlantic coast.

On our return journey we had reached St. Lo in Northern France, when Nicky said, "Stop, stop." She had seen a magnificent French cart with "À vendre" on the side. When I asked the owner "Combien?" she signaled, "What do you think?" I suggested £15.00 (equivalent to US$42 in 1965) in francs. The cart was ours! So we towed this huge farm cart, which towered over the Land Rover, with steel wheels over cobbled streets with chickens and dogs running for cover all the way back to the airport … where they said they could not get it on to the plane. So we had to drive to Cherbourg to catch a ferry. The French customs were fine (Had we "manged" the cheval?), whereas the British customs, typically with no sense of humor, spent hours working out what we would have to pay. Eventually they found that I should pay 20 percent—£3.00—not a disaster!

October 1965: A rare photo of Nicky and Oscar with the French farm cart we bought on our honeymoon in St. Lo, France. We towed it back to England behind our Land Rover (author's collection).

9

The Mill and Nim's Arrival (1966)

Before getting married I had moved into No. 3 Frognal Gardens, a semi-basement flat just off Church Row in Hampstead Village. The current tenant was a publisher, Ed Victor, who worked with Dux Schneider at Weidenfeld & Nicolson. Ed had moved to a flat on the floor above, enabling me to get an extremely good, small flat in a great location. Back from our holiday in France, Nicky took up permanent residence, and she obtained a job with interiors managing a Casa Pupo china shop in Hampstead Village. We bought a black Labrador puppy—Oscar. We decided it would be fun to try to find an old property which we could renovate as a weekend cottage. One weekend when we were down in the country Nicky spotted a derelict water mill in the Lambourn Valley at Weston. We discovered that many people had tried to buy the old mill, but the owners would not sell as the fishing was worth more than the property. So I went to see the agent in Ramsbury and pointed out to him that, as a structural engineer, I could see that the mill building was not safe. Children from the village could have a serious accident, in which case the owner would be liable. For my trump card I said that they could keep the fishing rights providing I owned the millstream under the house, the two-acre island and the distance to the center line of the river. He agreed and asked me to name my offer. As I would have to get a mortgage, I offered £2,000 (US$5,585 at 1966 exchange rates) or £2,500. To my amazement he agreed but suggested the figure should be £2,000 as I had to get a mortgage! This was early 1966. I managed to get a mortgage for £4,500, which included £2,500 to help me rebuild the entire building. As the river came through the center of the house, I installed a thick concrete ground slab with retaining walls on both sides of the river and the huge mill wheel. It took Nicky and me 18 months to complete all the work as we did everything ourselves except for the internal plumbing and electrics.

While we worked on the mill every weekend, during the week we enjoyed living in Hampstead. At the Everyman Cinema[1] we saw great classical films and modern French, Italian Left Bank classics like *L'Année dernière à Marienbad*.[2] While I was working on the construction of Northwick Park Hospital, Nicky was trying to sell very ornate china pyramids of fruit! Sometime in May 1966 Nicky was staying at Burlyns because she felt unwell, and I went up to London for the day by train. On returning Nicky met me at the station in a state of great excitement.

1966: The converted mill from the south, showing the water going under the house through the large wheel (*House & Garden*, May 1969, courtesy Conde Nast).

She had been to see her doctor that afternoon ... she was pregnant—against all predictions! What joy—how fantastic—something we had never planned or dared hope for!

I found that to pay my mortgage we would have to live at the mill and give up our lovely London flat, and I would have to commute from Newbury. As I needed more time for the construction of our new home, Nicky's parents very kindly said we could stay with them during Nicky's pregnancy. This would give us both more time, me for building and Nicky for painting and decorating, as the mill was only about 20 minutes drive from Burlyns. We worked very hard. I remember I was bricklaying with my radio on when England won the World Cup.

But the greatest event of that year was at University College Hospital, Euston Road, when Nicky gave birth to our lovely daughter Nim. Because the hospital was both excited and nervous about the importance of this unexpected birth, it had Nicky admitted for a week before her due date.

It was on a Saturday when Nicky telephoned her mother to tell her that she was beginning and, of course, I was working on the mill. Bridget telephoned the Weston Village Shop, who sent someone to find me. I drove my Land Rover back to Burlyns, quickly changed, drove to Reading (this was before the M4 was open from Newbury to Reading), took a train to Paddington and then a taxi to University College Hospital. As I walked into Nicky's private room, the nurse said,

"You have just arrived in time to see them both." I said "both," thinking, *Surely we didn't have twins?*

She then said, "No your wife and your lovely daughter!"

Nicky's reaction was "never again!"

Nicky and Nim stayed in hospital for several days, and Nicky's surgeon, who had removed her one ovary and injected the other, came to see her with champagne.

As Nim grew, Nicky became extremely thin and then started having jerks, for instance, dropping a fried egg on the floor as she tried to put it onto a plate for my breakfast. We moved into the mill, but the jerks became worse until one morning she had a full-blown epileptic fit. This was horrendous, Nicky lying on the floor jerking like mad, frothing at the mouth with an awful scream coming from inside her! After many tests, including a lumbar puncture, she was diagnosed with epilepsy.

Because I had to leave early every morning and did not return until about 7:30 in the evening and because Nicky was not meant to drive, I had to arrange help for us. Somehow we managed, and Nicky got well again, taking an enormous number of pills. Nim fortunately was a very easy child and came everywhere with us.

Although we stayed quite a lot at Burlyns, we did not give up our London flat until we could move into the mill, which was not until Nim was about one year old. As an example of how well Nim behaved, I remember one evening at Frognal Gardens, when Dux and Monique had returned from Turkey. We were all sitting on the floor drinking and smoking, as we all did in 1966, with Nim also on the floor in a Moses basket. Monique had just learned that she was pregnant, and Dux, as usual, was making a lot of noise telling a story which required him to thump the floor. To his amazement Nim slept through it all! "I really hope our child will be like Nim," he said. Tamara came about six months later!

December 17, 1966: Nicky and Nigel with Naomi (known as "Nim") (author's collection).

10

Thompson's Eating House (1970–1972)

In 1968 we designed and built a vertical extension to the cottage end of the mill, and in 1969 Nicky had another little girl, Sadie. She had reddish hair and freckles.

While we were still living at the mill, we bought two pet rabbits for Nim and Sadie which, initially, they looked after with great enthusiasm. One Saturday our neighbors, two young men from London—Stephen and Roger—called in to see our new rabbits. They owned Marsh Cottage where they kept their pets, even though they only came down at weekend because Roger was a theater producer and Stephen a hairdresser for the stars and pop world. People like Cliff Richard

1969: Nicky, Nigel, Nim and Sadie (author's collection).

10. Thompson's Eating House

and Jim Dale were frequent visitors. On this occasion, to Nim's delight, they brought their very large pet rabbit Thumper over to meet Mopsy and Flopsy. It all went so well until the next morning. Nim came in to see us and said Thumper had gone all stiff! He had. It was clearly rigor mortis in coitus interruptus. Sadly Nim and Sadie wheeled the dead Thumper surrounded in a bed of wild flowers back to Marsh Cottage. A few weeks later when moving the rabbit hutch, many little rabbits ran in all directions!

Nicky was always coming up with different ideas for schemes of what we could do next. Late in 1969 Nicky said we ought to open a bistro[1] in Hungerford.

1970: Thompson's Eating House–external view (author's collection).

At the time everyone claimed that there was nowhere locally to have a good meal out except for a few pubs like the Dundas Arms at Kintbury or The Bell at Ramsbury. We did further research which confirmed Nicky's opinion. I said I was prepared to do this providing we found a suitable place. We would have to live over the shop and manage the restaurant ourselves; it never works if you are relying on someone to manage it for you ... but we would need a chef. Nicky found an old pet food shop in the center of Hungerford which we could buy for £10,000 (US$23,962 at 1970 exchange rates), and we persuaded the sous chef from The Bell to join us. We visited a hugely successful new bistro which had just opened in London—Julies—all Victoriana, large mirrors, aspidistras, Bentwood chairs and scrubbed tables—ideal. We decided to call ourselves "Thompson's Eating House!" Nicky took herself off for training at a friend's pub to pick up tips on how to take table orders and record on a pin board, etc., etc. We designed the layout, had tables made, bought secondhand kitchen equipment from trade magazines, bought old gas lamps and bric-a-brac from Bermondsey market at 4:00 a.m., refurbished the whole place and built loos and what was to be our flat over the shop.

In the summer there was a steam fair held on Hungerford Common with all the ancient tractors and early steam merry-go-rounds and organs. Nicky persuaded me that we should take our French farm cart, towed by our skewbald farm horse, around the fair selling doughnuts (donuts) to advertise "Thompson's Eating House." The day was a huge success, the restaurant was looking terrific and we sent out invitations for a grand opening in September.

A week later we suffered the most terrible tragedy. Sadie, aged eleven months, managed to hang herself trying to climb out of her cot. We were absolutely shattered. Somehow we carried on. Little Sadie was buried in the churchyard of East Woodhay church where Nicky and I were married. Much later her grandmother, grandfather and Nicky's half brother Richard were also buried there.

Through hard work Thompson's became a huge success. After twelve months we were in two *Good Food Guides* and consequently were often frequented by politicians and even royalty. Virtually everything was made on the premises, the menu changed every day, and all vegetables, meat and fish were fresh. On one occasion, when Princess Anne's Lady-in-Waiting telephoned to make a reservation, Nicky told her to piss off as she thought it was me telephoning from London to wind her up! Fortunately that must have happened before, as the Lady-in-Waiting persisted, Nicky apologized and Princess Anne came!

11

Grove Farm and Sam's Arrival (1972)

After 18 months Nicky was pregnant again, and we agreed to try to find a spectacular old house, preferably derelict. Nicky found us Grove Farmhouse at Stitchcombe. It was a beautiful old Yeoman farmhouse which had a Queen Anne frontage tied on to the original house in about 1714. The house was south-facing, clearly empty and shuttered up. This was February 1972. Nicky, as usual, did her research and found that the owner had died but had previously given an option to buy the farm, 200 acres and the farmhouse to Geoffrey Young, who was the farmer managing the farm. Nicky also found out from the village shop that he did not want the farmhouse and was terrified that someone would come in and buy the whole farm … he did not know how to proceed, and he was an old bachelor who liked to play bridge with titled ladies!

With this information I asked Lady Jameson, the mother-in-law of Humphrey Wood, an architect friend of ours, to introduce me. I consequently telephoned Geoffrey and said I wanted to see him to discuss a proposition. With a bottle of Famous Grouse I arrived at Geoffrey's door and spent an evening talking about mutual friends, Lady Jameson and the Jellicoes, drinking the whisky and explaining that we only wanted the farmhouse. I told him that I had a very good lawyer who would help Geoffrey buy the whole farm, and we would at the time of signature buy the house from him. I said we could get the whole purchase carried out within one week. On price I said I would give him £2,000 more than the best price he thought he might get if he went to auction. He thought it might go for about £25,000—so we shook hands on £27,000 (US$67,318 at 1972 exchange rates). We completed seven days later on the farmhouse, outbuildings and 13 acres. Geoffrey had the farmland for £27,000, the total transaction being for £54,000.

My solicitor said it was the fastest house purchase he had ever carried out, and he thought one of the most successful. One year later Geoffrey Young sold the 200-acre farmland to John Burrows for £202,000!

During my early years of marriage, I spent a lot of my time at Arup in becoming the firm's hospital design specialist. After Northwick Park I was appointed the structural engineer for a new District General Hospital at York,[1] which again was designed for adaptability and change, and also in London for a new wing at the National Hospital for Nervous Diseases.[2] Most of this hospital work was with the architect firm Llewelyn-Davies Weeks. During the early '70s, I also worked on my first shopping center at Warrington (the Golden Square Shopping Centre, opened 1974) for a large developer who had, in partnership with the local authority, bought a run-down shopping area with housing over the shops and a fine old English pub, The Barley Mow. The project set new standards of quality and was highly thought of at the time. But the insurance companies which were helping to finance the whole scheme insisted that there should be no housing in the development and, for security reasons, that the whole center should be locked up at night. This was common practice for all new shopping centers in the '70s. Removing housing, which provided the life in the area, was, I believe, another serious mistake made by town planners during this period. This was recognized by the '80s and '90s, and future developments would nearly all contain mixed use of shopping, housing and leisure.

At around this time I was also doing some structural engineering for the architects David Levitt and David Bernstein[3] on housing restoration for Circle 33.[4] One day my colleagues at Arup were impressed when a number of well-known actors led by Ian McKellen came to the office to interview me alongside Levitt and Bernstein as their possible designers for Theatre 69, the new theater to be installed in the Corn Exchange in Manchester, now known as the Royal Exchange Theatre. This was a competition which I was very excited to win. As we were not able to put any loads on to the floor of the existing hall, I came up with a solution to hang the whole theater from two large trusses spanning between the four large brick columns supporting the Corn Exchange's huge dome. We wanted to expose the steel structure unprotected, which was contrary to fire regulations. Fortunately, Peter Rice had just had a similar challenge with unprotected steelwork for Centre Beauburg (now known as Centre Pompidou) in Paris. He had solved that problem with the aid of Margaret Law at the Fire Research Station (FRS). Margaret had now agreed to join Arup and set up a Fire Engineering Group. So, with Margaret's help, we were able to demonstrate that the steelwork for our theater needed no special protection, other than sprinklers, because there was sufficient volume within the hall and dome for smoke—so there would be time for everyone to be evacuated safely. On one occasion during the design process, I invited Ove Arup to join the architects and myself for a design meeting. I knew that David Levitt and David Bernstein would enjoy meeting Ove and that he, in turn, would be pleased to discuss and review a project. The Royal Exchange Theatre is a very exciting performance venue which was opened in 1976 by Sir Laurence Olivier.

After leaving the Lambourn Valley in 1970 for Hungerford, we had to think about schooling for Nim, who was then four. We managed to get her into Chilton Foliat Primary School even though the school was in Wiltshire, the neighboring county. This was a huge advantage because we did not have to make any changes when we moved 18 months later into Stitchcombe. While we were running Thompson's Eating House, we were open six days a week, even including Christmas Day. This was a huge success for older, lonely people as we had one huge table at which I carved the turkey, and Nim was running around helping!

11. Grove Farm and Sam's Arrival

1969–74: Theatre '69, Manchester, later known as the Royal Exchange Theatre, showing one of the two main trusses spanning between the original large brick columns. The entire theater is carried on these two steel trusses (Arup).

During that time at the restaurant Nim learned to write her first word. Unsurprisingly it was "Total" copied from the bottom of everyone's bill!

When the Federation of Rhodesia and Nyasaland broke up in 1962 my brother stayed on in Rhodesia helping the government avoid sanctions, and my mother and stepfather then moved to Malawi. Five years later my stepfather retired and with my mother moved down to Fish Hoek near Cape Town in South Africa. In the Spring of 1972 mother and Dennis came over to England for a holiday. Towards the end of this visit I had a very nice lunch with Mother in a little Italian restaurant in London near my office. She was very happy that Nim was doing well and that Nicky was pregnant again. Having also stayed with us in the middle of Grove Cottages while we were working on Grove House, Mother was now very happy that I had married Nicky and was very impressed with Nicky's huge energy and ability. She was also convinced that we would have a little boy. Mother was not, however, looking forward to going back to South Africa and would have preferred to stay in England. I saw her off into a taxi for her to go to Victoria Station and then down to Dover, where Dennis was staying. Later that afternoon I had a phone call from Dennis. He had met Mother off the train—she said she was not feeling well and then collapsed. She was now in hospital on a life support machine, having a blood clot on the brain.

I drove down to Dover the following morning. Mother looked terrible, totally different to the person I had lunch with the day before. She never regained consciousness, and I had to agree with Dennis that they should switch off the machine. She was only 60—I was 34. It was a huge shock to both Peter and me. Dennis went back to South Africa, sold the house, packed up and moved back to England. But he only lived for another two years.

Nicky, Nim and I were living in Grove House Cottages on the main road, which I had hired from Geoffrey Young, while we carried out a major refurbishment on our new house. We had, in fact, bought Grove House without even having the opportunity of inspecting the inside of the house because I knew that, whatever the condition, we would be able to fix it. We had bought the house because of its amazing position and its very elegant appearance! Once inside we realized we would have to carry out some major replanning because, on entering the front door, you were in total darkness as, at the back of the house, the roof went down to the ground with the staircase in the middle blocking out any light and inhibiting circulation between rooms. I realized I needed help, so I asked my good friend Brian Perry, the architect (married to Karin—Ove Arup's youngest daughter) to come to my office at Arup one evening to sketch out alternative solutions. After working for about two hours Brian suddenly said, "I think the best solution is to do this," and explained an idea which meant demolishing the staircase and building a new open staircase in a glass tower at the back of the house. This was brilliant and completely unlocked all the circulation difficulties. So whilst obtaining planning approval for our new scheme, we were able to get on with replacing the roof and treating the dry rot which had affected much of the ground floor.

Sam was born on September 8, 1972. Another great day. Nicky had been huge as Sam had been in no hurry to come into the world. He was a large baby, nearly 10 lb., with no hair and a big head. He was very strong and would try to lift himself up in the cot. The nurses were a bit nervous about this and took him away for examination!

Nicky developed her epilepsy again. Because I had to commute up to London, we hired a girl—Bridget Bretherton—to help us with Sam and drive Nicky around.

12

Iran and Phoebe's Arrival (1974–1976)

International oil price increases in the early 1970s led to large flows of money into Iran. As a result, in the 1970s, Iran achieved one of the highest growth rates anywhere in the world. Public expectations of ensuing benefits were high. The shah committed three years of petrodollar income to finance internal development in the country. U.S. and European consultants and contractors started queuing up to offer their services. Among other things, the shah announced a US$5 billion scheme to build a two square mile "city within a city" in Tehran. This would help realize his dream of making Iran's capital a major world city.

The period of 1973 and 1974 was the start of the Middle Eastern oil boom, so work in the building industry in the United Kingdom dried up. The Labour government had been building hospitals, schools and universities but put all construction on hold as they had run out of money. The private sector, which had been building shopping centers, also slowed down. It was time to take action. One of my colleagues at Arup had won a major project in Saudi Arabia with two German architects, Frei Otto[1] and Rolf Gutbrod.[2] I realized we had to look where the work was, and I thought of Iran.[3] Keeping my ear to the ground, I heard that Lord Llewellyn-Davies, the hospital architect, and Lord Jellicoe, Chairman of Warburgs and Tate & Lyle, were planning to visit Iran at the invitation of the shah[4] who wanted four teams from different countries each to build five new hospitals. The shah had invited the American, British, French and German governments to provide teams to compete for the commissions. Having worked with Lord Llewellyn-Davies and having met Lord Jellicoe through my daughter Nim being friendly with his daughter Emma at Chilton Foliat School, I joined forces with the two eminent Lords to be one of the groups competing to be selected by Her Majesty's Government as the British Hospital Group.

With Arup's blessing I flew to Tehran first class with Lord Llewellyn-Davies and Lord Jellicoe. George, 2nd Earl Jellicoe, KBE, DSO, MC, FRS had been a page boy at the coronation of King George VI. He was a distinguished soldier, second-in-command of the Special Boat Squadron and commander of the SAS during the war, an eminent diplomat in Washington, secretary general of the Baghdad pact, and more recently a politician in Edward Heath's shadow cabinet. He resigned from politics in 1973 when a scandal broke with his name along with that of Lord Lambton's in a call girl's diary. However, in 1973 Lord Jellicoe then concentrated on building a business career.

Through Lord Jellicoe's connections and the British Ambassador, we met senior politicians in the Shah's cabinet and also the Shah's fixer—Sir Shapur Reporter, who had received a knighthood for his efforts in persuading the shah to be a purchaser of Concorde! While the Lords Jellicoe and Llewellyn-Davies held further political discussions, I traveled by bus to Tabriz and Resayeh to inspect two of our possible hospital sites up near the border with Turkey and Azerbaijan. There were originally four British teams competing to build four hospitals. The Iranians only wanted one British team, and Her Majesty's Government refused to nominate only one team. Two teams were strong politically: George Jellicoe's team supported by LDW and Arup with Cementation as contractors; Chris Chataway's[5] team was supported by the architects Yorke Rosenberg and Mardell with Taylor Woodrow as contractors. I helped integrate the two teams together with both architects but with only one engineering company—Arup. We formed such a strong group that the other two UK teams withdrew from the competition.

Having been selected as the British hospital team to design, construct and equip four large general hospitals in Tabriz, Rezayeh, Mashad and further south at Ardebil in the desert, we now had to put teams together in the UK and arrange appropriate site inspections. At Arup Dr. Duncan Michael had already completed a formidable monument for the shah, the Shahyad Monument,[6] near the airport, the result of a competition won with two young Iranian Architects—Hossein Amanat and Manouchir Iranpour. The shah had the monument built to

1974: The Shahyad Monument in Tehran, built to celebrate the Shah's dynasty, was later renamed Azadi (Freedom) Tower by the Ayatollah (Arup).

12. Iran and Phoebe's Arrival 55

commemorate the 2,500th anniversary of the Persian Empire founded by Cyrus II the Great. When the Ayatollah[7] took over in 1979, it was renamed Azadi (Freedom) Tower.

While we were planning the hospitals, LDW were approached by the mayor of Tehran to prepare a new master plan for Abbas-Abad or Shahestan Pahlavi, a new city center for a redeveloped Tehran. With the help of Michael Lewis[8] I managed to get Arup also appointed by the mayor as the civil engineers for the design of all roads, bridges and transportation systems and traffic engineering for this great new city! We would be working with Jaquelin T. Robertson,[9] an American planner recruited by LDW, to lead the town and architectural planning team. The core of this new plan was the creation of a massive urban square to be known as Shah and Nation Square, which would match Tiananmen Square in Beijing.

The future Queen Noor of Jordan was at the time a single 24 year old (who I knew as Lisa Halaby) working in the LDW team in Tehran. Queen Noor reflected on her time in Tehran in the mid–1970s and wrote a beautifully succinct description of Shahestan Pahlavi:

> The new city center, with its views of the snowcapped Alborz mountain range, was to feature pedestrian malls, theaters, moving sidewalks, shopping galleries, shaded arcades, and terraced gardens. Government ministry buildings and foreign embassies would rim one of the largest open public spaces in the world, to be called Shah and Nation Square. The scale was monumental: Shah and Nation Square was designed to be larger than Red Square in Moscow, and the Shahanshah Boulevard, the broad, tree-lined avenue through the project's center, was intended to emulate the Champs-Elysées in Paris. My work as a

From left: Nigel Thompson, Chris Glaister (deputy to Jac Robertson), Jacquelin Robertson (leading planner for Llewelyn-Davies Weeks), Shah of Iran Mohammad Reza Pahlavi, Queen and Empress Farah Pahlavi, Nikpay (mayor of Tehran), 1976 (author's collection).

planning assistant entailed surveying and mapping all the buildings in the vast area surrounding the site.[10]

At that time most of Arup's UK offices had little work. I volunteered to go out to Iran and open an Arup office in order to obtain projects where much of the design work and all the drawings could be carried out in the UK—thereby reducing the number of engineers we would have to make redundant. I negotiated a very good salary with Arup, found a suitable office building in Tehran with accommodation above where visiting engineers could stay, found a property for Nicky, Nim and Sam, let our beautiful house at Stitchcombe to a visiting American banker and took my family to Iran.

Life was very busy in Iran. I was joined by a small team of very good, young engineers—Martin Manning[11] and Holger Koch-Nielsen.[12] We landed many projects, the office was always full of visiting engineers and we made huge use of the telex machine as, at that time, faxes had not yet taken over.

Having successfully completed the Shahyad Monument, and because we were already working on Alvar Aalto's[13] Museum of Modern Arts[14] in Shiraz, Arup was developing an enviable reputation in working creatively with architects in Iran. William Pereira[15] from California became a frequent visitor to my office as did Philip Johnson[16] and also I.M.Pei,[17] who had been encouraged by the shah to visit Tehran for a possible competition to design a new museum in Tehran for His Imperial Majesty. In

Shahestan Pahlavi: The new city center was to feature pedestrian malls, theaters, moving sidewalks, shopping galleries, shaded arcades and terraced gardens (courtesy Llewelyn Davies).

fact, it was from working successfully with American architects in Iran, and subsequently in the Gulf States, that some of our U.S. architect friends persuaded us to come to America and open an office.[18]

Nicky used to provide the team with lunches of shepherd's pie and lasagna. We had an apartment up in Shemiran, north of Tehran at the foothills of the Alborz mountains, where all the wealthy lived and the British Embassy had their summer retreat escaping from the intense summer heat of central Tehran. The traffic in Tehran was crazy. When winter arrived, the roads iced up, and all the cars skidded into one another. Whereas I drove our Land Rover, which some friends of ours had driven out to Tehran for us, Nicky drove a Deux Chevaux—Citroen II—which became much battered by the Tehran traffic. Both Nim and Sam went to school. Nim attended the British School which, when she started, had 60 children but, with the oil boom, within the two years grew to 600. Unsurprisingly the head had a nervous breakdown, and Nim, aged eight, sat at the back of a large classroom and learned nothing. Meanwhile, Sam, aged two, learned to sing Frère Jacques in ten languages, including French, Farsi, Urdu, Italian, German, English, Dutch and Spanish.

For the office we landed many projects. From the Shahyad architect Hossein Amanat we gained the commission for a new military academy for the Shah. From Manouchir Iranpour, who was now architect to the Shahbanou (the Queen), we were appointed for a number of museums and art galleries. Negaristan,[19] one of the largest such projects, was to be built underground in a huge hole! We were also appointed for another Tehran hospital, a new parliament building to be built as a huge pyramid and a tank factory for the Shah. These projects kept Arup engineers busy in London, Birmingham and Edinburgh.

Sardar Afkhami was the architect for the New Parliament Building, and he asked us to do the structural design. Mike Brown led a team of engineers volunteered from our Edinburgh Office and they completed the scheme design. Unbeknown to us, the project was completed after I left Iran. I discovered this when I saw on television a very senior Cleric making a speech from their New Parliament Building.

Nicky became pregnant again. She visited a good, local hospital where she had to sit handcuffed to the chair so that she could be examined. Unsurprisingly, Nicky told me that in no way would she have our child in Iran. So after 15 months we came back to England for Christmas. Nicky was to stay with Sam in the UK after Christmas to have the child while Nim would come back with me to Tehran. The baby was due in February. Just before Christmas Nicky became ill with Listeria monocytogenes, and Phoebe was born two months prematurely. Whereas Nicky quickly recovered, Phoebe developed meningitis followed by encephalitis with her head swelling. Unfortunately, surgeons would not be able to operate until February as they said they had to wait for the full nine months! I do not think they expected Phoebe to live through Christmas. Although very small she showed she was clearly a fighter and hung on with a very swollen head until, in February, Phoebe had her first "shunt" fitted by Mr. Adams at the Radcliffe Infirmary. So Nicky stayed in England with Phoebe and Sam while I went back to

Tehran with Nim and Moira as home help. Fortunately, I had to fly back and forth quite often.

On one occasion I was invited to lunch on a Thursday by the senior directors of Tate & Lyle, who were returning to London that afternoon by private jet. They invited me to join them. I duly collected Nim from school, rushed to the airport and boarded their plane where the four of us were looked after with champagne and caviar by a hostess. We stopped for refueling at Ankara and Geneva and then landed at Heathrow at about 1:00 a.m. Nim and I took a taxi to Burlyns. No one was expecting us, and we managed to get upstairs and into a spare bed in Nicky's bedroom without any of them waking up! What a joy and great surprise it was in the morning. I later learned that Tate & Lyle had to leave Iran as they were in a big dispute with the shah's government.

We had many good times in Iran. We explored the amazing country from Tehran up through the Alborz mountains and through the tea-growing plantations to the Caspian Sea. We traveled then from east to west along the Caspian, where Russia became the neighbor, Azerbaijan at the west and the Mongolian steppes in the East. We explored where the Kurds lived from Turkey to Iraq and into the salt desert of the South.

When visiting Mashad with Lord Jellicoe, the shah had arranged for us to visit the holy shrine in the wonderful Mashad Mosque. We were given a police guard. I will never forget the experience of joining hundreds of weeping pilgrims as they entered the most holy of Muslim shrines. The floors were covered with beautiful Persian carpets, but they were all soaking wet from the sweat and tears of the pilgrims. As we had to remove our shoes on entering, our socks soaked up the pilgrim's tears!

There were many special events that I remember from my two-and-a-half plus years running the Arup Iran office. There was a great Russian restaurant which we used to frequent where they served us caviar and blinis with a bottle of vodka encased in a lump of ice. In the winter on Fridays (their Saturdays) we used to drive up into the mountains and ski. The slopes were fully equipped with ski lifts and associated facilities, so it was here that we taught Nim to ski. There was also a special event when the shah laid the foundation stone for his new city, and the mayor, Mr. Nikpay, presented me to the Shah. At that time the Tehran office of LDW was being run by the American planner Jaqueline Taylor Robertson. At the office I also met Richard Helms,[20] the American ambassador, former head of the CIA during Richard Nixon's presidency, and Ron Ziegler,[21] Nixon's former press secretary, both being exiles from Watergate!

On another occasion I took my chairman Sir Jack Zunz to meet the British ambassador Anthony Parsons (later Sir Anthony Parsons, who helped Margaret Thatcher over the Falklands). Jack asked how the ambassador and the British government viewed the future of Iran. The ambassador elegantly replied that conventional wisdom was that the shah might succeed in turning Iran into a second world country ... providing the Persians didn't fuck it up, which was more than likely. Jack was shocked that such a well-spoken, articulate English patrician should use such language!

12. Iran and Phoebe's Arrival

During our last six months in Iran, I used to take Nim, aged eight, for riding lessons at the Tehran Country Club where wealthy Persians kept their stallions. No Iranian man could ride anything other than a stallion. So Nim had the experience of learning on strong, fiery, Arab stallions. She later told me she was always terrified but kept going to please me!

Arup has always been a family firm and demonstrated this in its concern for my family situation. Arrangements were made for Richard Haryott[22] to take over the running of the Tehran office from me and, towards the end of 1976, Nim and I returned to the United Kingdom. Our family was reunited, and we took back Grove House from our American banker tenants.

13

Doha and Kafrawi's Tower of Winds (1976–1978)

Back in London I was made a director both of Ove Arup & Partners International and Arup UK I was asked to take over Ted Happold's[1] work in the Arabian Gulf. Ted had left Arup to start his own consultancy practice in Bath, named Buro Happold.[2] At the time he had also been made a professor of the University of Bath. I inherited a project that was in an enormous mess. The project was a huge university in Doha, Qatar, for both men and women. With the exception of science, all the different faculties were duplicated so that the sexes would be separated. However, as science laboratories are expensive, the architect decided that with skillful time management the laboratories could be shared by placing them between the male and female faculties. The architect was Egyptian, Dr. Kamal El Kafrawi.[3] The client was the Amir Sheikh Khalifa bin Hamid Althani. The project was overseen by Mr. Hisham Qadumi, a very tough Palestinian who was in charge of the Amir's private office.

The project was in a hopeless mess. The team, which included the architects Renton Howard Wood Levin (RHWL) who we had employed as subcontractors, had completed detailed design drawings on a site very close to the ring road of Doha. On my first visit to Doha while I was sitting in the Amir's palace waiting to talk to our client about his nonpayment of fees, Arup's David Atling was checking out the electrical systems near the site when he was shot at; a bullet passed through the cab of the Land Rover he was driving. Prior to this event we had arranged for a post and wire fence to be erected around the whole of site No. 1. As David was retreating in haste, he witnessed men bulldozing the recently erected fence. When I reported this to Mr. Qadumi, he explained that the local Sheikh claimed the site was his and not the ruler's. He said not to worry because it wouldn't happen again and he'd get the fence reinstated. The fence was re-erected, and in the night huge lengths were again ripped out. At that stage we had stupidly spent nearly £1 million (US$1.8 million at 1976 exchange rates) without receiving any fees.

The Arup team under Ted Happold was used to working in the United Kingdom where they would do the work, send in a bill and get paid. In fact, they had left it to the finance office to send in the bill. We now had a situation where our client had suddenly decided that he had to change the site to several miles further out into the desert, and he was not going to pay as all the drawings related to a different site with completely different contours. Dr. Kafrawi, whom I had not yet met, had disappeared to Paris and was suing us for breach of copyright because our subcontractor, RHWL, had advertised in some promotional literature that they were the architects for the university. Arup's Fraser Anderson had flown down from the firm's Edinburgh office to take over Ted's work, which included the work in Saudi Arabia and Doha University.

On the University, Fraser had asked the Arup senior partner Bob Hobbs to help him try to deal with Dr. Kafrawi. Both in turn asked Ove Arup himself to get involved to try to smooth out difficulties, because now the architects RHWL were suing us for wrongful dismissal. It was at that stage that Fraser had bowed out and the senior partners asked me to take over the project. Because of my successful experience of working with difficult people in the Middle East, they put all the problems into my lap. The partners feared that Arup might end up losing £2 million on this project. We currently had about ten engineers on the Arup team. I realized that to deal with the issues I had first to get close to Dr. Kafrawi and sort out being sued for breach of copyright.

I flew to Paris and spent days getting to know Dr. Kamal el Kafrawi. My experience of working in Iran, my knowledge of Palestinian issues and my patience won him around after several good, long Parisian meals, many evenings with glasses of Johnny Walker Black Label and cigars! I had decided that the right thing to do was to sign an agreement giving him total copyright of all our designs. The Arup lawyer strongly advised against my proposed action. I learned a few years later that the lawyer had in fact written to our chairman saying I was doing the wrong thing and was completely mistaken in my approach. Fortunately, Roger Rigby, who was the Arup partnership secretary, deliberately put the letter in a drawer and failed to inform the chairman (or maybe he showed the chairman, and they agreed to back me!).

Anyway, I managed to get all lawsuits put away and, after a few years and many trips back and forth to Doha, we obtained payment of our fees, rebuilt the team, redesigned the project for a new site and constructed an amazing project in the desert.[4] I was supported by very good engineers, and my right-hand man was Mike Brown, a Scotsman who came backwards and forwards with me to Doha and Paris on numerous occasions. Mike helped manage our whole team, and together we handled a number of extremely difficult and nerve-wracking situations. David Anderson, Neil Noble, Tom Barker and Peter Gill were engineers who all contributed enormously to the success of this exciting project.

We had 20 engineers working in London and about ten permanently in Doha, supervising both the precast concrete factory,[5] which was built just for casting the panels for the buildings, and for checking all the work on the main site. Thank goodness the fax machine had now arrived, superseding the telex machine. The main contractor was Japanese, and the principal subcontractor was Hyundai, Korean. On the appointment of the subcontractor I said to the senior Japanese engineer, "Presumably you are happy having a Korean sub-contractor?"

To my surprise he said, "No!" He then explained that it was all due to their rice culture; the Japanese were farmers and had to work with their neighbors to manage the water, whereas the Koreans were hunters and would capture their water! The Japanese, who were always correct in abiding by local laws, had great difficulty in obtaining alcohol, whilst their Korean subcontractor had large quantities of whisky within days of establishing themselves in Qatar.

Mr. Hisham Qadumi, who ran the Amir's office, was a very difficult person to deal with as he tried to bully everyone and always had many reasons why he did not need to sanction payment of fees due! He would tell me to come to his office "bukra" (tomorrow) or after tomorrow "bath bukra." He would deliberately make us wait knowing that we probably needed to catch a plane back to the UK. I had to stay there and wait patiently. During 1977 and 1978 I spent around 200 days and nights each year in Qatar or traveling there.

Nicky Thompson and Mike Brown at Qatar University site, in front of typical classroom block with tower of wind (author's collection).

On one occasion when I was required to fly out for an important meeting with Mr. Qadumi, I was being driven to Heathrow as the snow started to fall. We were listening to Terry Wogan[6] on the radio, and I heard him saying that he hoped it would stop snowing as he was taking the inaugural flight of Concorde the next day to Singapore. Anyway, I arrived at the airport and eventually boarded the Gulf Air plane. We sat on the runway for seven hours waiting for de-icers. Gulf Air took us off the plane, put us into a hotel for the night and said hopefully we would fly the next day. The following day everything was in chaos again at Heathrow, and it didn't look as if we would fly. On the off chance I went across to the Concorde counter because I thought if any plane was going to take off, it would be Concorde. There was one returned seat. The first stop would be Bahrain, where I could get a connection to Doha, so I took the available seat. It was absolutely brilliant—an amazing experience not to be forgotten. We flew across Europe at 35,000 ft. and a speed of 450 mph. When we got to the Adriatic, we climbed to

13. Doha and Kafrawi's Tower of Winds

From left: Mike Brown, Dr. Kamal el Kafrawi, Nigel Thompson (author's collection).

Top left: laboratory blocks. Foreground: standard classroom blocks; each classroom is an octagon supporting a wind tower which naturally ventilates/cools the classroom (John Donat).

70,000 ft. and went through the sound barrier flying at twice the speed of sound. We landed in Bahrain three and a half hours later after a long lunch.

Towards the end of the university's construction the Crown Prince was so excited with the design of the university based upon the traditional "Wind Towers"[7] of Doha that they submitted the design for the Aga Khan Award at the Biennale in Venice. So Mike and I, accompanied by Dr. Kamal el Kafrawi and one of the princes, flew to Venice where we stayed, of course, in the famous Hotel Danieli. It was October, and Venice looked fantastic with the mists in the morning covering the low sun. It was my second visit to Venice, and I said to myself, *I must bring Nicky sometime.*

Following on from my experience in Iran where I had learned how to use the British Embassy, I always visited the ambassador and kept him informed on progress or lack of progress in getting our fees paid. On one occasion during the visit to Qatar by Douglas Hurd, UK foreign secretary, in the royal yacht I persuaded the ambassador not only to invite members of the Qatari government and British businessmen, but specifically Hisham Qaddumi, the technical adviser to the Amir, our client, to a party on the yacht. Subsequently, in the meeting of Douglas Hurd with the Amir, the problem of late payment of fees was discussed.

By the end of the project we had managed to be paid fees of over £13 million (US$25 million at 1978 exchange rates) with a very healthy profit for Arup. And not the £2 million loss envisaged! Arup's senior partners had said to me when I took over the project that if I could avoid making a loss, then I could keep 10 percent of any profit. Naturally I knew this was a joke and needless to say, this conversation was conveniently forgotten by them. My final flight out to Doha was for the university opening ceremony, to be conducted by His Highness the Amir on February 23, 1985.[8]

Being away so often and not knowing when I would be able to return was a strain on Nicky, who was coping with three children. The youngest, Phoebe, was so ill that from time to time she had to have replacement shunts fitted.

On one occasion when I had only just managed to return for Christmas Day, we were invited out by our neighbors, the Evans family in Stitchcombe, for a Boxing Day black-tie evening whist drive and dinner. Partners were split up, there were perhaps 28 people there around seven tables, and you were required to move from table to table between each hand of cards. The evening had been going fairly well for both Nicky and me—with us both winning—until we came to the same table but as opponents. I had a very good hand, and after playing halfway through I could see that I could not help winning all the remaining tricks in the hand. So I put my cards down on the table, saying, "All the rest are mine" (as you often do when playing bridge)!

Whereupon Nicky exploded, saying, "You fucking cheat!" and with her hands knocked all the cards off the table. There was total silence in the room.

Our host stood up and said, "I think it is now time for us to have a break for dinner!" Which we all did and returned to cards after dinner. Nicky, understandably, received all the sympathy. It was an event that has not been forgotten by those who were present!

13. Doha and Kafrawi's Tower of Winds

Because of my success in Iran, and subsequently at the university in Doha, my group was appointed for further work in the Arabian Gulf: the Gulf University in Bahrain and other projects in Doha, Abu Dhabi, Dubai and later a design and build for a new hospital for Sharjah, United Arab Emirates. There, I appointed Llewelyn Davies and Weeks to be our architect subcontractor. We also obtained projects in England because we now had a conservative government with Margaret Thatcher.

In 1978, with the arrival of the Ayatollah in Iran, the Tehran Arup office had to close. The British Hospital Group who had carried out the initial agreement with the Iranians on the design work for the four hospitals had, of course, not been paid. So the group, led by Cementation International and Taylor Woodrow, proceeded to sue the Iranians and threatened to serve a writ on one of the Iranian airlines. As I said earlier, after a protracted period the Iranians paid up!

14

Life in Wiltshire and Africa Again (1977)

Back in Wiltshire we often saw the Jellicoes, and on one occasion Nicky and I were invited to lunch at Tidcombe Manor because Sir Anthony Parsons, formerly the British ambassador to Iran, and his wife were also invited. We were enjoying eating caviar when George suddenly said, "Philippa, where's the other tin? You know, the one that Hoveida gave me" (Hoveyda had been the Prime Minister under the Shah).

Philippa made a face and said, "It looked as if it had gone bad, so I gave it to the cat!" It was the best golden caviar—the only kind the shah ate!

After leaving Iran we had seen that Hoveida[1], along with our client the mayor of Tehran, Nikpay,[2] had both been executed along with many others by the Ayatollah's Red Guard!

Nim was now at boarding school, and Sam was at Greenwood, the pre-prep school in a part of Burlyns where Nicky's parents lived. Phoebe was now doing much better, and Nicky was establishing herself as a classical singer, continually being invited to sing in concerts as a mezzo-soprano soloist—on two occasions with the Bach choir in Oxford and also in Malmesbury Abbey. She was also in demand for many other events such as weddings and funerals.

We were enjoying living at Grove. We had a pig, Matilda, whom a neighbor had won at Axford fete and could no longer manage. The children were delighted and after taking her to the boar many weeks later, we had 14 "Matiglets"! We had three litters of piglets, always a large number, and they would frequently escape, which the children loved as it was a great game catching the speedy piglets. On one occasion one of our piglets started to swell up enormously. Nicky asked one of our local farmers what might be the cause, and his reply was, "Have you checked whether it has an arsehole?" He was right—it didn't! So we got the farmer to come to Grove to help me, acting as surgeon. Having measured the distance below the tail on other piglets, I bathed the proposed area with alcohol and then with a scalpel made a significant cut and squeezed. It worked! So I bathed the wound and neatly stitched the new orifice. The piglet lived happily until his time was up. Once they were large enough Nicky used to take them off to the abattoir and then sell the prepared piglets on the school run to other parents. We also had a horse and some ponies which village children used to come and ride and take to pony shows.

14. Life in Wiltshire and Africa Again

I haven't yet mentioned that in 1976, the hot summer after I returned from Iran, Nicky had the great idea to start a vineyard. At that time I was traveling much more around the world to California, working with architects, and to Australia, South Africa and Zimbabwe, as I had taken over the Arup responsibility for coordination with the Arup African practices in Southern Africa. So Nicky and I were able to extend some of my Arup visits to inspect vineyards in the Napa Valley, San Francisco, and the vineyards of Southern Australia and South Africa. We planted our vineyard in 1977. The vines, which we imported from Germany, were grafted on to Californian root stock to safeguard against the disease phylloxera. In 1977 we planted three varieties on a seven-acre field:

- Müller-Thurgau—approximately 1,000 vines. It makes an excellent wine with a slight Muscat flavor.
- Reichensteiner—approximately 1,000 vines. The wine has a full-bodied, flowery character.
- Siegerrebe—approximately 1,000 vines. The wine has a powerful flavor and bouquet and blends well with Müller-Thurgau and Reichensteiner.

The vines grew so well and the wine proved to be excellent so that, after six years, in 1983, we expanded by building four polytunnels into which we planted 1,000 vines of Bacchus and 240 vines of Scheurebe, all for blending with the original vines. Because some grew in polytunnels, they produced enhanced sugar naturally. The site was ideal, south-facing and sloping down towards the River Kennet. The soil was similar to the champagne country—loam overlying chalk.

I was very busy at Arup leading my team and on the weekends, when I was not traveling, we worked in the vineyard. Because there was potentially a lot of work on the vineyard, I obtained planning permission to turn an old cow barn into two cottages for vineyard workers. I also obtained planning to build some additional agricultural buildings behind our house to store the wine vats and make the wine.

One day when Nicky and I were working outside getting an old cowshed ready to convert into a cottage, Sam—aged about four—came to us and said, "Phoebe's gone all funny and blue." We managed to get a doctor, who then travelled in an ambulance with Phoebe to the John Radcliffe Hospital in Oxford. With a new shunt Phoebe was fine, but it was only because Sam saved her life. Sam was quite accident-prone at that time. On one occasion we had to take him to the local hospital to remove a marble he had managed to push up and get lodged in one nostril. A few weeks later, we were again at the hospital because he had managed to get one of his fingers nearly bitten off by a horse. But the most serious incident was also that month when in the kitchen Nicky, whilst making some coffee, had just put boiling water into a mug on the worktop when Sam had come up behind her and reached up for the mug—spilling the boiling water over his shoulder! Thankfully it was not on his face. Obviously he was rushed up to hospital once again to be treated for some very serious burns. Someone at the hospital suggested that Nicky would probably be reported to Social Welfare! Sam carries some very impressive manly scars.

Grove House, vineyard cottage, the granary, barns and the eastern half of the vineyard as seen from Stitchcombe Hill (author's collection).

Nicky also decided that we should turn the old granary into a shop—but it was on the wrong side of the house, some 13 ft. (about 4 m) west of our house. Again we obtained planning permission to turn the granary into a shop but then decided to move it to the other side of our house—some 33 ft. (10 m) east of Grove House. We decided to make the move two days before Christmas. I would send a letter (held up in the Christmas post!) explaining that owing to tree route damage, we had to move the granary, and I would reapply for planning in the New Year. We took the roof off the granary to reduce the weight, put a steel cage below the timber frame and hired a crane, which then lifted the whole structure onto a long farm trailer. The plan was to take the trailer down our drive to the main road, turn the tractor and bring it up a second drive, which we had constructed the other side of the house. While we were moving the whole structure, we would take the 16 staddle stones[3] around to the new position ready to receive the well-braced timber structure. All went well until the tires on the trailer burst when we were on the main road. In no time we were joined by the police because we were blocking the main road between Marlborough and Ramsbury.

Anyway, somehow we completed the repositioning of our granary shop and obtained planning for the new position. Sometime later we received notification that our house and granary were listed Grade II with the granary in its original position!

14. Life in Wiltshire and Africa Again

Granary placed on to tractor trailer (author's collection).

At Arup, in addition to the overseas work, there was a resurgence of activity in England and, for my group particularly, in refurbishment of shopping centers, new offices (with many existing buildings keeping their previous Edwardian facade) and theaters. In particular there was refurbishment of The Old Vic and a new Nottingham Concert Hall. In the office, computers were gradually taking over the role of the draughtsman, the reinforcement detailer and—to the great relief of many—the activity of preparing bending schedules. Also, of course, the computer was now being used by the engineers for analysis and the preparation of the required calculations. It was for the design of the Sydney Opera House that Arup first used the computer for analysis and the definition of the complex geometry of the structure—back in 1960.

Nicky and I were also invited by Lord Jellicoe to accompany them with some other leading businessmen from the city, banking and commerce on a business marketing trip to Thailand and the Philippines.

In Thailand we visited some of the Tate & Lyle projects, and in the Philippines we met President Ferdinand Marcos[4] and had dinner with his wife Imelda.[5] The purpose of the mission was to identify possible international projects and, in my case, for Arup to make contacts with leading international bankers and contractors. It was shortly after my visit to the Philippines that our Hong Kong office opened a design office to help with the huge load they had in Hong Kong.

At this time I had taken over from Geoffrey Wood (one of the original Arup partners) his position as Arup representative on the London Chamber of

Commerce and Industry (LCCI),[6] Lord Jellicoe was president of the LCCI from 1980 to 1983. It was during that time at an official function that Lord Jellicoe first presented me to Her Majesty The Queen, where she remarked upon the Arup design for the elegant new curved bridge at Kylesku,[7] Sutherland.

In 1980 I was asked to lead an LCCI mission to the newly independent countries of Zambia and Zimbabwe. It was marvellous for me to revisit the part of Africa I knew so well and had left 20 years previously in 1960. As we got off the plane in Lusaka, the great smell of the soil and sun of Africa hit me. It was a feeling of homecoming. Arup still had a small office in Lusaka, and I was able to give us greater profile. As leader of the mission I was invited to take my team to dinner with President Kenneth Kaunda at Government House. It was a very successful evening, and I had an extremely interesting discussion with the president. During dinner, he offered me South African wine. When I queried this saying, "Surely you should be boycotting all South African produce," he replied that you have to be pragmatic—this wine is cheaper! I declined the wine and chose to have the pomegranate juice he was drinking. I also decided not to tell him about our meeting in 1959 when I, as a Rhodesian soldier, had locked him in prison. Our visit was so successful that it enabled improved relations between the president and the British high commission.

After Zambia we visited Zimbabwe, where I appeared on their television news, which was interesting. Some friends who I had not seen or had contact with in 20 years saw me on television and came to see me at our hotel. We also saw President Mugabe and, again, I was able to increase the profile of the Arup office. Colin Moynihan, later Lord Moynihan, from Tate & Lyle, was on the mission with me and we enjoyed playing golf on my old course at Ruwa Country Club. He became a very good friend. He had won a silver medal in the Olympics held in Moscow as a Cox for the British Eight. He later became Minister for Sport in John Major's government. After the mission, before returning to the UK, I flew from Zimbabwe to Johannesburg to stay with my brother Peter, whom I had not seen for 20 years, and to meet Sally and their three lovely daughters Sarah, Frances and Jude. It was wonderful meeting them all, and we decided that over the next Christmas I should bring Nicky, Nim, Sam and Phoebe out to Africa for a holiday.

About one year later we all flew out to Johannesburg, where I picked up a car from Cliff McMillan at the Arup South African office, and—with Peter, Sally and the three girls—we drove in convoy from Jo'burg up to Zimbabwe, repeating a drive Peter and I had made 25 years previously. When across the border into Zimbabwe, we noticed that the game was much more evident as, clearly, it had not been hunted during the long war of independence. We stopped at the same small hotel just over the border where we had stayed so many years before and saw a giraffe drinking out of what had been a swimming pool.

We drove on up through Bulawayo to Victoria Falls—what a most wonderful sight—and stayed at a campsite on the banks of the Zambezi just above the falls. We bought steak at a local shop and cooked it outside while monkeys were in the trees around us and warthogs ran through the undergrowth. The Africans were exceptionally friendly as they were not used to visitors and the tourist trade

14. Life in Wiltshire and Africa Again

From left: (Back row includes) Peter Thompson, Sally Thompson, Mike Hill, Nigel Thompson, Nicky Thompson; (Front row includes) Peter & Sally's children Jude, Sarah and Frances, Mike & Ann's son and daughter, Nigel & Nicky's children Phoebe, Nim and Sam (Photograph by Ann Hill; author's collection).

had not yet begun. Phoebe was entranced by the babies strapped to their mothers' backs, and she referred to them as chocolate babies as she was not used to seeing black children. The Matabele people of that area and the majority Shona of Zimbabwe were all wonderful, friendly people.

After visiting the falls, we spent the next three days in Wankie (now Hwange) Game Reserve where we stayed in two rondavels.[8] The game in Wankie was abundant, and we saw many elephants, lions, buffaloes, zebras, giraffes etc. We had Christmas in the camp. I cooked and carved a turkey, and Sally was doing the Christmas pudding. There was suddenly the most enormous explosion. Sally had put the tinned Christmas pudding in water to boil on a stove in one of the rondavels, forgetting to pierce the lid of the tin! The tin had gone straight through the ceiling into our roof!

After Zimbabwe we drove down to the Indian Ocean in South Africa, around Swaziland to Ladysmith, then to the sandy beaches of Margate before returning to Jo'burg and then back to London. It was the most wonderful holiday and a great way for our children to meet their cousins and to introduce my family to Africa.

15

Commercial Buildings in the 1980s

Recession in the United Kingdom in the middle of the 1970s was followed by the country's biggest construction boom for 100 years. London was the "hot spot." London's commercial buildings' surge would lead to rebuilding much of the City of London, redeveloping parts of the West End and regenerating the Docklands to the east of the city. There were also significant developments of London's airports and pioneering science and business parks. As a firm, Arup had extensive knowledge of ground conditions and existing buildings throughout London, gained from working continuously in the city since 1946. We were one of the first consulting engineers to have our own specialist geotechnical engineering consultancy, so we were well placed to benefit from the construction boom. For many

Wigmore Street in the West End of London—the old Debenham and Freebody Store (Arup).

15. Commercial Buildings in the 1980s

commercial projects there would be planning requirements that existing façades and other architectural features should be retained.

A typical example is the old Debenham and Freebody Department Store in Wigmore Street, built in 1907. Now in 1981 London and Leeds Estates proposed its redevelopment as fully air-conditioned office space, retail shopping and residential development. This would involve working within constraints imposed by the need to retain the entire façade and other historical features of the listed building. Of particular significance were the highly figured and decorated faience façade[1] and a magnificent marble-clad entrance hall with grand staircase and decorative ceilings. I had a very good multidisciplinary team at Arup working with me, including Mike Taylor, Mark Facer and Malcolm Barrie. We produced a scheme which held up the façade and main entrance hall and demolished the remaining building, which we then replaced with a new lightweight steel structure. The new structure was supported by a basement raft with augured piles under each column to minimize differential settlement. The new building had to be air-conditioned, which was relatively easy to install for the first two levels because of generous ceiling heights. At the higher levels we had to use raised floors as well as the ceiling voids for air distribution and other services.

Wigmore Street—facade supported by temporary steel structure during demolition of store interior (Arup).

By contrast, Aviation House is an example of new build of a different kind of city scale and location, being at an international airport (London Gatwick). The four-story, 16,000 m² (more than 172,000 ft²) Aviation House is the headquarters of the Civil Aviation Authority's Safety Regulation Group. Again I used a multi-disciplinary team of structural, mechanical and electrical engineers—Mike Taylor, Mark Facer and John Haddon. This was a 19-month fast-track scheme built as two separate wings around a central atrium. The wings are connected horizontally by link bridges at each level and vertically by glass-enclosed elevators from ground level to any bridge.

16

Theaters and The Old Vic (1982)

I was responsible for the engineering design for a lot of theaters and concert halls, both new builds and refurbishments. They were always very difficult and complex for the structural and air-conditioning designers because there are few right angles. All floors generally sloped, galleries were curved, and ceiling heights provided insufficient space into which we could insert structure and ducts for air-conditioning and other services.

My first theater was Manchester Royal Exchange with architects Levitt & Bernstein (Chapter Eleven). Projects ranged in size from The Wilde Theatre at South Hill Park Arts Centre in Bracknell,[1] again with Levitt & Bernstein, to the New Nottingham Concert Hall[2] with Nick Thompson at architects RHWL (Renton, Howard, Wood, Levin). There were designs and redesigns for the Syrian National Theatre, Damascus—an international competition we won with Nick Thompson. There were proposals for the redevelopment of the London South Bank with Terry Farrell. Then there was the Players' Theatre under Embankment Place with David Binns of Sandy Brown Associates.

But the most famous was The Old Vic in 1982, where my team was John Pilkington and Alan Foster. The Old Vic has an intriguing history. It opened in 1818, the year after Waterloo Bridge opened, as the Royal Coburg (architect Rudolf Cabarel), with a seating capacity of 4,000. In 1833 the theater was redecorated and renamed The Royal Victoria Hall. It quickly became known as The Old Vic, with a reputation as a bawdy and drunken music hall. In 1869 the auditorium was completely rebuilt (architect J.T. Robinson), and the seating capacity was reduced to 2,800. In 1880 The Old Vic was leased to the social reformer Emma Cons, who turned it into The Royal Victoria Coffee House, run on temperance lines offering lectures and concerts.

In 1912 Lilian Baylis took over, and two years later The Old Vic Shakespeare Company was formed under the direction of Ben Greet. Rebuilding and alteration works were carried out in 1925–26 (architect Frank Matcham). In 1931 Sadlers Wells was opened by The Old Vic under Lilian Baylis' management. Between 1931 and 1936 Shakespeare, opera and ballet alternated between The Old Vic and Sadlers Wells.

In 1941 the theater was bomb-damaged. Renovations were undertaken, including the design and erection of a new proscenium, and the theater was reopened in 1950. The Old Vic Company was disbanded in 1963, but the theater was by now one of London's premier venues. It was a natural choice to house the fledgling National Theatre, which took a lease on the building and appointed Laurence Olivier director. Again extensive alterations were carried out. In 1976 the National Theatre moved to its new home on the South Bank and the Arts Council withdrew its subsidy to The Old Vic. The resident company went into liquidation, and the governors sold the freehold to the Canadian businessman Ed Mirvish[3] in August 1982.

Ed Mirvish decided to restore the theater. Nick Thompson of RHWL, supported by my team, won the competition to restore the theater to its late-nineteenth-century form and decoration with extensive front of house and back. It was a very difficult project as it had been modified on many occasions, often in unsatisfactory ways and mostly using timber. The entrance foyer was gutted and completely refurbished. The auditorium, seating capacity 1,077, is now entered through new corridors between the foyer and gangways down either side of the auditorium. The Lilian Baylis Circle is accessible through new vomitory entrances which bring the audience in at the front of the Circle rather than at the very back of the auditorium.

We had to modify the layout of the stalls with increased seating capacity, the dress circle with new entrances and newly stepped seating and the gallery with wider seatways to accommodate proper seats. We extended the stage to the full width of the site. The auditorium had to be air-conditioned and its acoustic

The Old Vic restored and refurbished (Arup).

16. Theaters and The Old Vic

The Old Vic new interiors (Arup).

performance increased to NR25.[4] The whole building had to be treated with fire protection systems as well as sprinklers. Modernizing the theater necessitated substantial structural alterations in all areas. Because of the complexity of the original structure, we had to fabricate most of the steelwork on site, and having no existing structural drawings made diagnosis of how the existing structure was meant to work more difficult. The Old Vic was a demanding but most exciting project.

Part II

A Leader in Building Engineering and Business Development

"Nigel and I worked together many years ago. It was a very different time. Professional firms worked primarily within their own national borders with the occasional international adventure. Business development was a relatively new concept viewed with wariness and suspicion. At that time, Nigel and I worked for separate firms that were teamed to pursue major international projects, and subsequently we worked together within Arup.

Looking back now and in retrospect, I recognize that Nigel was a major catalyst for change within Arup. Nigel's internationalism was consequential in moving the firm from a UK-centric engineering practice focused on supporting high-minded architects into a modern, multi-functional, international engineering firm. Surely there were other factors as well, but Nigel's role was significant.

Nigel had lived abroad growing up and had an early interest in the international marketplace. He successfully pursued major international infrastructure projects and geographic markets of opportunity. He opened international offices and helped to build an Arup presence in new markets worldwide. His resolute international focus became an important factor in the arch of Arup's development.

Nigel was never conventional. He never colored between the lines. He upset the apple cart of the status-quo again and again, but, in so doing, he nudged traditionalist Arup toward a larger role in international markets that helped position the firm for the future. Perhaps it would have gotten there without him, but it would have likely taken much longer. Nigel ruffled feathers in the conservative professionalism of that time which operated largely on the belief that work was derived solely through engineering excellence. Nigel was always clear-eyed about the objective and went for it.

Years have passed and times have changed since Nigel was involved at Arup. The firm has grown enormously in size, in geographic locations, and in engineering disciplines. But Nigel was in Arup at a pivotal time of change in the world and in the architecture and engineering marketplace. Nigel's early recognition of this change and this new reality had a measurable and everlasting impact at the time and on Arup's growth going forward."

—Anne (Kriken) Mann

"When I resigned my commission in the Canadian Army after ten years' service, I was an experienced military man and a rookie in everything else.

I desperately needed a mentor, more so than I realized at the time.

By a stroke of good fortune I got the best possible mentor—Nigel Thompson.

That happened because I joined Arup, known for its excellence in engineering and commitment to cutting edge design ... but traditional in many ways.

Nigel recognized that the traditional approaches were not going to be enough as the world was moving faster and faster.

Nigel respected the culture of Arup but, with the dawning of the new digital era, drove forward new approaches to winning business that delivered layers of sophistication and, at the same time, reduced the cost of winning business.

Nigel's vision, quiet tenacity and respect for tradition were key to the next chapter of Arup.... I watched, learned and was hugely privileged to be a part of it.

I'm certain that Nigel has no idea how BIG a role in my life he played. His guidance and care (plus a few kicks in the ass) set me up for the great life I have now.

Nigel—you risked on me when I really did not deserve your confidence. Thank you—I am so grateful."

—Tim Haig

17

Joining the Arup Main Board (1984)

Ove Arup Partnership had been run firstly by Ove and then by Peter Dunican. Between them they'd steered the partnership forward for nearly 40 years. Main board reorganization in 1984 and the retirement of Peter Dunican set in place new cochairmen and expanded membership from nine to 17. I was one of the eight newcomers. The original nine were: Sir Ove Arup, Peter Dunican, Jack Zunz, Ron Hobbs, Povl Ahm, Sir Philip Dowson, Michael Lewis, John Martin and Duncan Michael assisted by Roger Rigby, partnership secretary and Stuart Irons, finance director. On the formation of the new board Roger Rigby retired and was replaced by Keith Dawson, partnership secretary, and Jenny Baster for legal matters.

By the early 1980s Arup had become a confusing organization to outsiders (and to a lot of insiders too). Principal letterheads were Ove Arup Partnership, Ove Arup & Partners (Consulting Engineers) and Arup Associates (architects, engineers and quantity surveyors). But clients could receive letters with all sorts of other headings relating to other operating parts of Ove Arup Partnership (OAP). None of this mattered in the slightest, of course, as long as the client was getting what he or she wanted. However, the overall firm's policy was vested in the OAP main board which, by the beginning of 1984, comprised nine people. Of these Sir Ove—he had been knighted in 1971—was coming into the office on just a couple of days a week to pursue his interests in design and engineering matters. He was a main board member but wasn't attending board meetings anymore. OAP had been reconstituted in 1977 with Peter Dunican as chairman. Arup Associates were represented on the main board by Ron Hobbs and Philip Dowson. Ove Arup & Partners civil divisions were represented by Povl Ahm and Michael (Mick) Lewis. Ove Arup & Partners buildings divisions were represented by Jack Zunz, John Martin and Duncan Michael. OAP board members also sat on the boards of appropriate operating parts. For example, Duncan Michael is a Scot with interest and experience in tall building design. So it was appropriate for him to be the OAP board member sitting on the boards of Ove Arup & Partners Scotland and Ove Arup & Partners Hong Kong Ltd.

All this was fine but couldn't go on. The reasons were that the firm was diversifying in terms of disciplines and was expanding in terms of staff and geographical spread. New fields of activity and new markets were underrepresented, or

82 Part II—A Leader in Building Engineering and Business Development

unrepresented, at main board level. Also, because there were only nine main board directors, there'd be logistical problems when any were away on business, on holiday or ill. Essentially, when members of the main board were absent, for whatever reason, the firm was run by their high-powered secretaries.

Existing main board members Jack Zunz and Ron Hobbs would be the cochairmen of the newly constituted board. Jack and Ron were both engineers. Jack was at the time a leader of Ove Arup & Partners, and Ron was a leader of Arup Associates. Being invited to join the main board at such a crossroads in the firm's development was a highlight of my time with Arup. In addition to running my very successful multidisciplinary group and leading on most of my projects, I'd been asked to be business development director for the partnership worldwide. My responsibilities would include the development of business strategies for obtaining building, civil and industrial engineering commissions. I'd also be taking the lead in training engineers in business and marketing skills to supplement their technical skills.

The huge honor of being asked to join the board can be measured to some

From left: (Back row) Peter Foggo, Stuart Irons, Anne Emmerson, Lillian Foggo, Duncan Michael, Pat Clayden, Joan Michael, Julie Martin, John Martin, Janet Shears, Mike Shears, Patricia Rigby, Nigel Thompson, Nicky Thompson, Virginia Haryott, Jorgen Nissen, Juliette Nissen, Sally Dowson, Philip Dowson, Mick Lewis, Richard Haryott; (Seated) Sylvia Rice, Zina Lewis, Babs Zunz, Ove Arup, Li Arup, Birgit Ahm, Maisie Hobbs; (Cross-legged) Jack Zunz, Bob Emmerson, Peter Rice, Bob Hobbs, Roger Rigby, Ken Clayden (Tomas Laski Ltd.; author's collection).

17. Joining the Arup Main Board

extent by the caliber of my fellow "new faces" in the "class of 1984": Ken Clayden, Peter Rice, Peter Foggo, Bob Emmerson, Mike Shears, Richard Haryott and Jorgen Nissen. Ken Claydon had been asked to represent Ove Arup & Partners regional UK offices. Ken had almost singlehandedly founded and built up Arup offices in Cardiff and then Bristol. He'd grown local workloads, won work from other more established local consultants and personally maintained close links between London, Cardiff and Bristol. On top of all that, he'd maintained commitments to firm-wide computing and to many other Arup day-to-day activities. Ken died suddenly in 1990 with a brain tumor.

Peter Rice relished challenges, seeking to advance engineering skills to realize the most ambitious of architectural concepts. Peter's collaborators came to learn that if anyone could make their ideas work, then it would be him. He was one of the original bright, young engineers on Sydney Opera House; then he along with Richard Rogers and Renzo Riano won Centre Beauburg (later known as Centre Pompidou) in Paris, then the Lloyd's Building in London, followed by Kansai International Airport in Japan. Peter died in 1992 at the age of 57, also with a brain tumor. He was already the leading engineer of his generation.

Peter Foggo was one of the leading architects of that same generation, and he would die in 1993, aged 63, with another brain tumor. His obituary in the *Architects' Journal*, July 14, 1993, began with, "Peter Foggo was an immensely gifted and engaging architect, a giant of the profession cut down in his prime...." Peter designed some of the City of London's finest commercial spaces but was a team player who didn't sign his name to buildings. Having done the first four phases of Broadgate, and when Arup Associates decided not to grow to take on later phases, Peter would leave Arup to set up his own practice, Foggo Associates.

Bob Emmerson had spent time in the States. Back in the UK he'd been the Arup director responsible for one of the biggest of the firm's early multidisciplinary building engineering successes—Triton Court in Finsbury Circus. This was, at the time, the largest building redevelopment in Western Europe.

Mike Shears had spent time in the States too. Then he'd taken over Arup's Industrial and Offshore Engineering (IOE) group at the time when the firm had low expectations of industrial facilities work and didn't believe it would ever make any money out of the North Sea. Mike proved everyone wrong. He built up a capability range which took in most aspects of offshore development, albeit not on a single project. On the strength of that track record, Mike and his team won an initial concrete gravity substructure (CGS) contract with Hamilton Brothers. This would lead to Arup becoming a leader in the architecture, management and technology of CGS design and delivery.

Richard Haryott had in the 1970s, when in his thirties, taken the lead on a colossal and hugely successful project—the National Exhibition Centre (NEC) at Birmingham. The NEC became the busiest exhibition center in Europe, staging more than 180 exhibitions each year. Richard took over the Iran office from me in 1976 when I returned to be based in the UK.

Jorgen Nissen, last but not least, brought bridge engineering design style to the main board. Jorgen would become one of the few engineers to join up two countries that had been physically separated since the Ice Age. His 1,624 m (5,328 ft.) Oresund Bridge, the second longest suspension bridge in the world, is part of a 17 km (10.5 mile) road and rail link which joins Sweden with the rest of Europe. This link cuts journey time between Malmo and Copenhagen from 45 minutes (ferry) to just over ten minutes (car). He was also responsible for the brilliantly elegantly curved road bridge at Kylesku, Sutherland, Scotland.

Arup was, in 1984, a bigger and more capable firm than it ever had been. But few people knew it. Mick Lewis was looking to create a corporate identity that would make the firm intelligible to outsiders. My interest at the time was the point at which communications joined up with business development.

Arup always won plenty of awards for its projects, but these were in construction industry competitions where everybody knew everybody. An exception was The Queen's Award for Technological Achievement in 1969 for Sydney Opera House, which had raised awareness for the firm within the country as a whole. But Sydney had been a long time ago, and I felt we needed to raise our profile again. So we applied for and were awarded The Queen's Award for Export Achievement in 1984. It helped, of course, that two of the United Kingdom's biggest overseas projects—the University of Qatar and Kansai International Airport in Japan—were at the time nearing completion.

In 1986 Sir Ove was elected an Honorary Royal Academician.[1] This would be the final major acknowledgment to a towering figure in the construction world. Ove died, aged 92, on February 5, 1988. Jack Zunz was knighted in 1989 and in the same year Peter Dunican died.

18

British Overseas Trade Board and the Vineyard (1983–1984)

In 1983 Lord Jellicoe, under Margaret Thatcher's government, was appointed chairman of the British Overseas Trade Board (BOTB), taking over from Lord Limerick[1] in promoting British exports, which was hard and not very rewarding work, particularly as the Foreign Office (FCO) and the Department for Trade and Industry (DTI) argue over the way work and money are divided. This is one of the longest-running bureaucratic turf battles in government. Jellicoe inherited the Duke of Kent[2] as his vice-chairman and, with Her Majesty's Government (HMG) approval, he invited me to join the BOTB. Jack Zunz, chairman of the Arup Board, was pleased that I was invited to join the BOTB because it helped increase the Arup profile both internationally and internally within the government, particularly as government ministers and secretaries of state frequently attended our BOTB meetings to discuss policy. I learned a lot from being on the board, and it was through the meetings that I first met Margaret Thatcher, whom I would meet again when I was promoting specialist engineering skills in an exhibition entitled *Better Made in Britain*[3] (I explain more about this work in a following chapter on "Business Development").

It was an extremely interesting time to be on the board as the Miners' Strike[4] (1984–1985) was reaching its climax, and Sir Ian Macgregor,[5] chairman of both British Steel and the National Coal Board, attended on a number of occasions. Macgregor had been chosen by Margaret Thatcher specifically to defeat the miners. After he had completed his mission, he stood down from the chairmanships. His successor Sir Robert Haslam[6] was to be appointed as chairman for both steel and coal and to similarly attend the BOTB (while initially remaining chairman of Tate and Lyall). Other frequent attendees at the BOTB included Michael Heseltine,[7] Kenneth Baker[8] and Lord Shackleton,[9] son of the great explorer,[10] who, although a Labour member of the House of Lords, was a great friend of Lord Jellicoe.

Members of the British Overseas Trade Board were always advised to contact the British embassy or the British High Commission when overseas on business to keep them informed on activities and to exchange intelligence. On one occasion, when the Arup Board of all the Arup Overseas Partnerships was to have its annual meeting in Singapore, I arranged for the British High Commission to hold

Nigel Thompson shakes hands with Ferdinand Marcos, President of the Philippines. Next to the President is George, 2nd Earl Jellicoe (author's collection).

an evening drinks party for Arup in the wonderful, old High Commission building, which had the most beautiful gardens right in the center of town. For the Singaporians it was always a very popular meeting place, so the party was very well attended by both government officials and clients. A business development success!

Back in England at Grove House our vineyard was now producing very good wine, providing the frost had not destroyed all the fruiting buds in May. We produced 14,000 bottles of Stitchcombe in 1983—one of our best years—all grown, picked, crushed and fermented at Grove. Then bottled, corked and individually labeled by Nicky with three labels—front, back and on the collar—all applied on the kitchen table using egg-white as glue! We were now selling the wine to Rules[11] restaurant in London. Because the wine was so good (it was a white, dry, light and slightly fruity summer wine), we had no difficulty in selling cases—even to some French friends who took bottles back to Paris. I think they enjoyed asking friends where the wine came from! Our wine was very similar to a Loire wine with some of the flavour from Alsace.

In 1983 we were delighted that Nim, now aged 16, managed to get into the 6th form at Marlborough College. She had done well at Stonar, recovering from her lack of education in Iran, obtaining some good "O" level GCEs and then passing a rigorous interview at the College. At that time Marlborough was only taking a few girls into the 6th form, and, as the girls would be significantly outnumbered by the boys, they had to be resilient and good-looking to survive!

Having Nim at the College was a help for the vineyard because she was able to bring a number of students from her class out to Grove to help in harvesting the grapes. Other pickers we used were gypsy women who were great regardless of the weather. They would still keep picking, even if raining, invariably with a cigarette in their mouths!

Sam was a boarder at Cheam from the age of eight. Cheam was the prep school that Prince Charles went to. He appeared to be enjoying school and particularly the rugby and cricket. Like Nim, he would sometimes come out for Sunday lunch, which was often interrupted by someone ringing the bell outside our shop. We would shout "customers," and Sam would often go out and serve. Although he was only ten, the customers were always amazed about Sam's knowledge of the wine—his patter would describe the grape varieties and explain why this year's wine still had a little "petience." He was our best salesman as he invariably sold a whole case!

Phoebe, who was now six, would help in the picking and particularly enjoyed the party atmosphere at our tastings.

19

Hospitals in the 1980s

St. Mary's, London, UK;
Onassis Hospital, Athens, Greece;
Yangon Hospital, Burma (Myanmar)

Hospital design in the 1980s was like airport and urban design. We were talking big buildings and associated infrastructure to support patients, residential staff, and all the visitors and employees living off-site. The size of facilities necessitated working with a multidisciplinary team of the highest order, with additional input from many specialist disciplines. Hospital design presented huge opportunities for engineers in transportation planning, pedestrian movement, noise and vibration analysis, communications, broadcasting and IT. New technologies had to be introduced into solutions that had to be responsive to future change. We had to do this while still meeting the complex, often contradictory, economic and political requirements.

Most hospitals are built in reinforced concrete. Construction in concrete is fast and cost-effective. Services can be distributed beneath flat soffits. Concrete performs well in terms of fire resistance, acoustics and vibration. It is robust, which helps confer adaptability for the future.

In the 1980s we were appointed by St. Mary's Hospital, Paddington, the major acute hospital for northwest London, as well as a maternity center with consultant and midwife-led services. This project demonstrated Llewellyn-Davies Weeks (LDW) working with me and my team as their "engineer of choice" and "one-stop-shop" for engineering design services. But I'd like to start the St. Mary's story way back, because it is characteristic of ways in which the expansion and redevelopment of Victorian hospitals in the UK has been driven by willpower, opportunity and acquisition rather than by comprehensive planning.

The foundation stone of the hospital was laid by Prince Albert (later to become Prince Consort) on June 28, 1845. That original building (now known as the Cambridge Wing) was designed by Thomas Hopper[1] in the style of a country house. It opened on June 13, 1851, with 150 beds for the sick poor, supported by charitable donations. St. Mary's was intended to be a teaching hospital, and its medical school was opened in 1854. Buildings continued to be added to the site, which is bisected by the world's first (1863) stretch of underground railway,[2] constructed by the "cut and cover" method directly under South Wharf Road.

West of site, in 1875, the Great Western Railway built a parcels depot and stables for 600 horses around a courtyard (these buildings were converted to hospital uses in the 1960s and are now Grade II listed). To the east, an outpatients' department and dispensary were opened in 1883. The Clarence Memorial Wing was built 1893–1904 to a design by Sir William Emerson[3] (this building is also Grade II listed). An additional floor of operating theaters and wards was built on to the original hospital in 1928. On South Wharf Road the Lindo Wing, for paying patients, was built in 1933–37 to a "neo-Georgian imperial" design by Sir Edwin Cooper,[4] as were new residences for nurses (Salton House) and a new medical school (now the Imperial College Medical School and Wright-Fleming Institute of Microbiology).

St. Mary's was clearly a very important center for health and medical research. It became famous globally in 1928 when one of the most momentous events in history took place here. It was at St. Mary's, in a small and dusty laboratory, where Alexander Fleming discovered penicillin when bacteria in a Petri dish became contaminated by a mysterious mold. The hospital attracted the world's media again when Prince William of Wales, heir to the British throne, was born in the hospital's Lindo Wing on June 21, 1982.

In the following year, 1983, the Acrow[5] group of engineering and construction companies went bankrupt. Its buildings next to the canal basin were acquired by St. Mary's, and this created a continuous hospital frontage on to the South Wharf Road. LDW and I had worked at St. Mary's before, but the early 1980s saw us designing a major new addition to the west of the site. This was the Queen Elizabeth the Queen Mother Wing, a ten-story general hospital building. The commission also included a large bridge structure. Construction commenced on site in 1984, and the building was opened on March 24, 1988.

I'd like to contrast this necessarily "piecemeal" approach to hospital facilities growth with the opportunity to design a brand-new hospital on a dedicated urban site. Aristotle Onassis was one of the most important shipowners of the twentieth century and the founder of Olympic Airways. His son Alexander wanted to fly airliners but because of myopia (nearsightedness) became a commercial pilot and guided taxi-airplanes. Alexander had in his vehicle store two small Piaggio 136 amphibian airplanes that he considered dangerous and referred to as "death traps" in conversation with his father. By the time that Aristotle came around to his son's wish to replace the aging Piaggios with helicopters, regular flyer Donald McGregor had gone down with an eye injury. Aristotle asked Alexander to train up a new pilot, Donald McClusker, to take the planes to Miami to be sold. On January 21, 1973, one of the Piaggios, with Alexander seated in front of Donald McClusker, took off on a training flight from the runway at Athens Airport. Some 30 cm (about 1 ft.) off the ground the aircraft rolled to the right, nose-dived into the ground, spun around in circles for 500 m (1640 ft.) and came to a halt with smashed nose, tail and left wing. When help arrived, Alexander was recognized only by the monogram on his handkerchief.

90   Part II—A Leader in Building Engineering and Business Development

St Mary's Hospital, Paddington, West London (Arup).

Aristotle Onassis rushed to the Athens hospital where his son lay deep in a coma with shattered face and irreversible brain damage. On the following day Alexander died, aged just 25. Aristotle Onassis was racked with guilt for not agreeing to substitute the Piaggios earlier, which had fatefully placed Alexander in the plane with an inexperienced pilot.

In 1975 Aristotle Onassis died, having bequeathed in his handwritten will half his fortune to his surviving child, his daughter Christina, and the other half for the creation of a private, independent Public Benefit Foundation, to be named after his son. His wish was that the foundation would perpetuate the Onassis spirit through charity, art and the development of Greece.

Among the most important public-benefit projects of the Onassis Foundation was the Onassis Cardiac Surgery Center. This is a 180-bed hospital for the diagnosis, treatment and rehabilitation of coronary disease, for the teaching of the staff involved and for heart research. It cost more than US$75,000,000 to build in the 1980s, on an 8,000 m² (more than 86,000 ft²) site on Syngro Avenue, Athens. Peter Stone,[6] Partner at Llewellyn-Davies Weeks, was chosen as architect to work in collaboration with the Greek architectural office of K Kapsambelis and K Stamatis. LDW wanted me and my team to be their engineer. Mike Brown, who was my number two on Doha, took on most of the work and visits for me. We designed the building to a revolutionary concept. All of the most vital operational units are concentrated on a single floor. This arrangement didn't exist in any of the world's notable cardiac surgery centers of the time but offered patients the prospect of faster, more effective treatment and recovery. The hospital was donated, fully operational, to the Greek state during an official ceremony held on October 6, 1992.

My team and I put together a lot of bids in the 1980s. Bids were costly to produce, but we were able to recycle data and our conversion rate for hospital projects was very high. The bids that interested me most were the ones that we did with LDW. This was because of our history of working well together and because the Arup range of engineering skills complemented perfectly LDW's hospital planning and architectural design capabilities.

Bob Trew[7] of LDW invited me to assist in making submission for New Yangon General Hospital in Burma. John Pascoe, Arup's technical writer, joined me and John Pilkington to assist in the World Bank submission. We had to convert our existing submission material to World Bank format and produce all the new data requirements to World Bank specifications. There was the usual number crunching to be done on fee proposals and attaching project values/consultants' fees to similar types of hospitals that we had designed and supervised in the past—a complicated process.

As my team was very busy I asked Brian Baxter and Roger Milburn of Arup's Manchester office to take on the project for me. They had young engineers with recent and current experience of working on a whole string of hospital developments in the northwest of England (and the World Bank liked youth in its project teams). The outcome was that we won the job, and LDW worked with Arup Manchester. Burma came under military rule in 1988 with the new regime changing the country's name to the Union of Myanmar in the following year. Arup did not wish to help legitimize the military government by continuing to work in Burma—we stayed on long enough to meet our commitment to Phase 1 of the hospital but no longer.

I've chosen to visit Paddington, Athens and Yangon—projects with the same architect—to demonstrate the variety of work going on in just one corner of my group's hospital business in the 1980s.

In 1981 I was invited by Ashok Tendle, the manager of the turnkey project contractor Eastern Limited, Sharjah, to lead the design for a new 300-bed general hospital for the Ministry of Public Works and Housing, Sharjah, in the United Arab Emirates. Unsurprisingly, I asked LDW to join us as architect but to be appointed directly by Eastern Limited for the architectural and health planning.

Design and contract documentation for this 35,000 m² (376,737 ft²), 300-bed general hospital were completed within 11 months of appointment of the turnkey contractor. Shaded courtyards were used to bring light deep into the interior of the building. Precast concrete was used extensively to provide shading elements and facilitate rapid construction.

The structural form is of cast insitu reinforced concrete trough floors with slid beam strips and columns. Columns are based on a 6.6 m (21.6 ft.) grid with a 3.3 m (10.8 ft.) slid strip through the spine of the building to form the "hospital street" at each level. The building is designed for seismic Zone 2.

The hospital is air-conditioned by an all-air constant volume reheat system with no recirculation in medical areas. A high standard of filtration is used throughout, and the critical operating theater and burns unit areas have ultra-high-efficiency, third-stage filters for 99.9 percent efficiency.

Our work at Sharjah was finally completed in 1988. Some of the engineers who worked with me on this project were Jim Keble-White, Ian Wattridge and Roy Smith. The manager of the contractor Eastern Limited, Ashok Tendle, was to resurface in 1990 as Terry Farrell's managing director.

We were, of course, working simultaneously on many other hospital projects with other architects. We were also involved, to a greater or lesser extent, on hospital projects being led from other Arup offices. The 1,100 bed Westmead Hospital in New South Wales, Australia, was completed in 1980. We got involved in designing a new generation of hospitals and healthcare facilities in California. The 137 m (450 ft.) tall, 27-story Queen Mary Hospital went up in Hong Kong (it's still the world's second tallest hospital building). The Prince Philip Dental Teaching Hospital was one of Hong Kong's first fully air-conditioned buildings (30,000 m² / 323,000 ft², 11-stories, 240 dental chairs). We continued to work on principal hospitals in southern Africa and West Africa, where Nigeria's new 540-bed Enugu hospital involved Arup in infrastructure design (roads and drainage) as well as structure.

Besides building works we got involved in the full range of hospital study-type projects, including building inspections, renovation proposals (e.g., over-cladding), fire and smoke safety studies, CFD analysis (computational fluid dynamics was a new skill I'd introduced into the firm to help us analyze the movement of people, air or smoke), acoustic studies, energy-use studies, seismic design studies, transportation planning and pedestrian movement analysis.

20

Shopping Centers and Retail Developments in the 1980s

In the middle of the '80s Mike Taylor and I made a campaign to get closer to the big shopping center clients and architects. We initially focused on upgrading and the refurbishment of tired shopping centers. We started to land a number of small projects which in turn led to getting to know more architects and clients like Capital & Counties, Legal & General, Norwich Union, Prudential, Ladbrokes and Friends Provident, all of whom owned large and often tired centers which needed serious improvements. During our marketing campaign, we went to a number of international conferences on shopping centers. A most notable one was held in Vienna where Mike and I, with our wives, joined a group of architects and clients to travel out and return on the Orient Express. It made a marvelous long weekend with a final black-tie dinner and dance in the Hofburg Imperial Palace in Vienna. I will never forget dancing the Viennese waltz in the great ballroom with Nicky in a magnificent dress I had bought for her that afternoon in Vienna's "Bond Street." We also, of course, visited the famous Spanish Riding School with its world famous Lipizzaners.[1]

Following on from the conference, we were recommended for two new shopping center developments by architect Nigel Warner of Chapman Taylor. One was a small development in Putney, just off the High Street. I also received a telephone call from the chairman of Estates and General, who asked me to come and see him to outline our experience not only on the design of shopping centers but also on digging huge holes and the economics of building a major two-story center underground. I convinced him of our considerable experience and technical expertise. But I emphasized that underground structures were generally far too expensive, and I frankly doubted if we would be able to make the economics work. Because, to keep the cost down as low as possible, structures had to be very simple and as lightweight as possible. He then explained to me their ideas of creating a center of more than 1,000,000 square feet in two and three levels below an existing car park next to the Castle on the Hill in the center of Norwich. Being next to the Castle would introduce the requirement for a significant archaeological investigation, another potentially huge expense of time and delay. In 1983 the project

Nicky and Nigel Thompson enjoy drinks aboard the Orient Express with their friends (not shown) Michael and Carole Taylor (photo by Michael Taylor).

began with the appointment of Lambert Scott and Innes, strong local architects, who obviously knew both Norwich City Council and the site extremely well. The plan was to replace the car park, which was crisscrossed by roads, with a green city center park with walkways, ornamental pools, a glazed pavilion[2] and boundary roads. It would be named Castle Mall. The mall's entrances would blend with Edwardian two-story and three-story buildings on the edge of the development. Below the surface would be a four-level retail center, services and car parking[3] extending across the six-acre (2.5 ha) site. Additional complications included difficult ground conditions and the need to build under key roads without closing them.

20. Shopping Centers and Retail Developments in the 1980s

Castle Mall Shopping Centre—continuous row of bored piles restrained with huge ground anchors. The huge hole in the ground extends to both sides, and beneath, the main road which leads to Norwich Castle, shown in foreground (Arup).

Five years of studies, planning inquiries and development negotiations ensued before the concept and development arrangements could be concluded. Development began as a collaboration between Friends Provident, Estates and General and Norwich City Council. Scheme design commenced in November 1988, and the management contractor Bovis was appointed in January 1989. During design development, the concept for the park was modified to respond to local authority opinion, pressure from local groups and budget restrictions. But the brief remained unchanged throughout—we were to provide a setting and an approach (both vehicular and pedestrian) to the castle, to create a gathering place relating to the upper level of the mall and to make an essentially traditional park.[4] To implement our proposals we had to excavate the site over its full plan area, requiring a retaining structure between 8 m (26 ft.) and 20 m (66 ft.) high around the full perimeter. The depth of excavation and constraints on ground movement led us to design a cast-in-place concrete retaining wall. Once again Malcolm Barrie was the structural engineer working with me who did most of the design.

The site was declared an ancient monument, and Friends Provident put up more than £600,000 (US$984,000 at 1989 exchange rates) to help fund a major archaeological dig. This enabled Norfolk Archaeology Unit to make the most of the year that was available to them, from the time just prior to perimeter pile wall construction start-up in September 1989. While the archaeologists were still completing their work, excavation began to an average 15 m (49 ft.) depth (350,000 cubic meters / 12,360,133 cubic feet of material were removed from the site). Ground anchors went in from June 1990—the ground anchors contract was the largest of its type in western Europe. Another major temporary work was a 77 m (253 ft.) long Bailey bridge which served as an independent approach to the castle, enabling more than 20,000 visitors per month to continue using the castle throughout the project.

The shopping center foundation works started in January 1991. The foundations are simple—generally pad footings but with 2.1 m (6.9 ft.) diameter piles where top-down construction was required for the two road crossings involved. However, we had to do an extensive site investigation, researching and analyzing the ground and the interaction of structure with ground. The pads are constructed on chalk containing solution fissures (areas where the chalk has been dissolved over time by water percolating down through the ground). To overcome this, we developed a process of compaction-grouting with a high-viscosity cementitious paste injected at high pressure (30 bar / 435 psi). We designed the piles using a new approach which delivered a significant reduction in the required length.

As a shopping center, Castle Mall was a significant project because it not only knitted together all the main shopping streets of Norwich city center but also opened at the time of a major shift in government policy relating to shopping, town centers and transport. In PPG (Planning Policy Guidance) 6 and PPG 13, the government made clear that out-of-town shopping centers which generated car-based journeys would no longer be generally desirable because of the associated road congestion and traffic pollution. It had also come to believe that out-of-town shopping centers led to the "doughnut (donut) syndrome"—the decay of urban fabric in inner-city areas. Instead, the government sought to redirect retailing activities back to town centers or at least back to edge-of-town locations.

Castle Mall opened to the public in September 1993. By April 1994 it had become the focal point for more than 200,000 shoppers each week. There were five major stores and 75 small units together with kiosks, specialty retailers, restaurants, cafés and gallery food court. In a MORI poll of the time, 83 percent of young, local people who were asked voted Castle Mall "the best-liked modern building in Norwich."

Another project came directly as a result of attending the shopping center conference. In 1987 I was telephoned by the development manager of Ladbroke Group Properties who had appointed The Halpen Group as architects, with whom we had previously carried out the successful rebuilding and restoration of Wigmore Street. He asked if I would take on a rebuilding/conversion and restoration of the BBC building in Portland Place at the top of Regents Street. They had bought the listed building off the BBC and wished to convert it into a top-class hotel. I was delighted to accept, and as it happens, I also knew the building well from studying it with Terry Farrell during our unsuccessful competition entry for the new BBC building. We did an excellent job in converting and restoring the BBC office building into a fine, top-class hotel. The project was completed in 1990 and the Arup engineers working with me were John Pilkington, Alan Foster, Chris Barber and Julian Olley.

21

Lutyens House (1986–1987)

In October 1986 automation of trading—the "Big Bang"—liberated the activities of the London Stock Exchange. This changed the ways we do business forever. The redevelopment of Britannic House, One Finsbury Circus, transformed one of London's finest steel framed buildings of the 1920s into prestigious offices with a building services and IT environment appropriate to commercial life after Big Bang.

City developer Greycoat had bought the freehold of Britannic House. In October 1985 they sought my advice on proposed redevelopment. It was a year before Big Bang but at a time when all commercial building planning was looking beyond 1986. This added enormous complexity to the challenge of redeveloping a landmark building that had been designed by Sir Edwin Lutyens[1] (we called the building "Lutyens House" for several years before it reverted to its original name on project completion). Greycoat knew that because of the sensitivity of the building and its location, only a transparently excellent solution would stand any chance of receiving the necessary consents. Because the architecture would have to be in sympathy with the Lutyens design, there would be even greater pressure on the engineering team to contribute vision, lateral thinking and technical brilliance.

The Britannic House story has its beginnings in the late eighteenth century, when George Dance the Younger laid out Finsbury Circus in its charismatic oval form. The center of one of the circles that Dance used to form the oval was taken by Lutyens as the starting point for the main axis which divides the Britannic House plot in two. Lutyens continued this axis through the building, in the form of the main entrance corridor, and cranked it at its western end to penetrate the building's Moorgate Street façade. Around this central axis Lutyens grouped all his main interiors. As completed in 1924, Britannic House had three major interior spaces—the staircase, straddling the axis and located on the fourth floor, the chairman's room (looking on to Finsbury Circus) and the double-height boardroom. The rest of the accommodation in the building was mundane, consisting mainly of cellular offices off long corridors lit either by windows to the street or by windows into light wells. Constraints on building redevelopment included the odd shape of the site, the relatively low ceiling height of the lower stories (awkward for location of the now mandatory under-floor services), the central location

Britannic House (Lutyens House), Finsbury Circus, City of London, after reconstruction and refurbishment (Arup).

of the stair and boardroom (neutralizing the central area of the building) and the need to light the deep plan.

The building had been Grade II listed in 1950. Planning consent for its redevelopment was conditional on retention of the building's ornate Portland Stone façades, its marble-lined ground floor entrance halls and its barrel-vaulted central staircases. Any proposed reconstruction necessitated detailed discussions on internal planning with the Corporation of London, English Heritage and the Lutyens Trust.

Greycoat wanted to thoroughly reorganize the interior, introducing an atrium and lifts and moving the boardroom. Their initial proposal, designed by William Nimmo and Partners, received planning approval from the Corporation but was refused listed building consent by English Heritage. Hoping to build on this partial success, Greycoat brought Peter Inskip and Peter Jenkins into the architectural team to help address the consents issue. Inskip and Jenkins began by studying the geometry governing Lutyens's design. They looked at contemporary Lutyens's designs of comparable buildings and investigated his sources. An intriguing idea began to emerge.

Lutyens had clearly used the circular geometry of Finsbury Circus as inspiration for the organization of the plan of Britannic House. Research by the two

Peters now revealed that a possible source for the design of the Lutyens elevation was the Palazzo Farnese Caprarola by Giacomo Vignola[2]—notable for its circular cortile. They discovered also that a Lutyens plan now in the RIBA drawings collection, and almost certainly an early design for Britannic House, included a circular light well. Had Lutyens, at one stage, considered following the plan, as well as the elevation, of the Vignola palazzo? The possibility appealed to Inskip and Jenkins. They worked with me and my team, led by Martin Gates-Sumner, to test this solution against the demands of both the building and the developer's brief. Placing the main floor of a circular atrium at first-floor level meant that most of the main Lutyens interior elements would be undisturbed, while shafts could be sunk to ground level to open up glimpses of the atrium space from the Lutyens axial corridor. The staircase could be contained within the atrium, while the chairman's room fell outside this area.

With a great deal of structural ingenuity, all the interiors were reconciled with the circular atrium except for the boardroom. But Inskip and Jenkins were determined to persevere with this solution because the atrium brought to the project big advantages of making sense of the challenging building plot and maximizing the amount of well-lit, open floorplan area. The boardroom was rebuilt in the basement, and its original location is marked by small projections from the atrium wall—a quirky feature to look out for it you should visit this building.

At the beginning of February 1987 the London Advisory Committee of English Heritage viewed the new Peter Inskip and Peter Jenkins design and declared itself happy. Conditions were drafted to ensure that the new design would be built as proposed (with William Nimmo as executive architect). The Lutyens Trust, which had opposed the initial scheme, wrote that it "would like to commend the architects for their concern to follow Lutyens's geometry and symmetry and the developers for revizing their approach to the building."

Already the scheme was being mooted as a potential model for the sensitive alteration and adaptation of historically important but obsolete commercial buildings. We had, however, much to achieve before Britannic House could become the role model that it deserved to be. Greycoat's Construction Director, Chris Strickland, in conversation with journalist Martin Spring, noted that "... Compared with Lutyens House, our other scheme to build offices over Charing Cross Station (see Chapter 23) is like a greenfield site! There, once you get above deck level, it's all about speed, efficiency and economy. But here it's like Swiss watch-making. There is such little space to manoeuver, and we have to think laterally all the time to come up with new techniques."

Taylor Woodrow Management Contracting was running the contract. Its project manager Paul Turrell, again in conversation with Martin Spring, commented "... My last job with Taylor Woodrow was Heathrow Terminal 4, where we had a construction programme network with 700 activities on a contract worth £200 million (US$327.4 million approximately at 1987 exchange rates). But this job is so intricate that we are using a network with a similar number of activities, though the contract is worth only £30 million (US$49 million approximately at 1987 exchange rates)."

In terms of physical construction, I'd reckoned that the biggest challenge to the development team would be posed by ground heave. Once Arup had been commissioned as consulting engineer for the project, we calculated that demolition of existing building would result in ground heave at the center of the site by as much as 20 mm (up to 0.8 in.), which could seriously affect the marble panels on the retained halls and staircase.

Our technical solution was to insert "hinges" into the existing structures that would flex if the ground heave were to become excessive. The hinges were achieved by temporarily removing the marble panels at critical points and deliberately weakening the brick backing walls by drilling vertical rows of holes through them. The idea was that the rows of holes in the brickwork would work like the perforations in a page of postage stamps. Additionally, we monitored ground heave throughout the construction period. Some 30 datum points around the retained ground floor were surveyed at fortnightly intervals by building surveyors Plowman Craven, who then plotted the movements for each point on a graph. Ian Scott, the Arup resident RE, estimated ground heave would be at its maximum on completion of the demolition work—a prediction that turned out to be correct. No movement cracks formed in any of the hinges drilled into the retained brick walls.

It was good that we were doing both the structural and M&E engineering because the job called for a closely coordinated approach (even better—we were also the geotechnical engineer, controls and commissioning engineer and fire safety designer). Because of the tightness of space on the site, we were also charged with responsibility for detailed coordination of subcontractors' drawings. Two of my engineers and the appropriate sub-contractors' draughtsmen worked together in the same office on their installation drawings. To optimize the available space within the building, we wanted to thread our service runs through the structural steel beams. So we designed special castellated steel beams. Because existing floor-to-floor heights had to be retained, we decided to retain the existing steel structure adjacent to the retained façades to a depth of 9 m (29.5 ft.) of the first bay. An additional reason to retain this structure was because Lutyens's heavily modeled stone façades included two setbacks on about 1 m (3.3 ft.) at third and sixth floor levels, which would otherwise have required enormous temporary shoring to counteract the eccentric loads. To enable service runs to be threaded through the ceiling voids of the retained bays, we simply cut holes through the webs of the existing beams (an approach consistent with our use of castellated beams for the new build!).

The redeveloped Britannic House was completed in 1989. The building was opened by HRH The Prince of Wales at his specific request. It was leased to British Petroleum and fitted out as their headquarters and is now the international headquarters of BP Amoco. The project is a model of urban redevelopment because all the servicing requirements of a modern office development were fitted into a volume never intended to accommodate them. Close coordination of structure and services enabled the introduction of an additional office floor within the height of the original building. We'd achieved more space—better space. We'd demonstrated that, with inspiration and multidisciplinary teamwork, historic buildings could be sympathetically recycled as efficient workplaces which retained their original "wow" factor. In doing this, we'd created the yardstick against which the work of future developers and their consultants would be measured.

22

Nicky's Singing

Nicky took this photograph of the family around the time that she was carrying out a number of singing concerts. I was busy in London building up a very strong multidisciplinary group—that is, mechanical, electrical, plumbing and structural engineers—specializing in commercial buildings. Nicky, at home, was busy marketing the vineyard labeling, selling and delivering cases of wine and singing semi-professionally. Her voice had matured into the most beautiful, deep mezzo-soprano. From 1977 right through the '80s she attended music festivals organized by the British Federation of Music Festivals and was, on numerous occasions, given awards of distinction. Three years running she was awarded distinction at the Oxford Music Festival for Solo Opera and Lieder. In 1978, 1979 and 1981 she won at the Wiltshire Festival for Lieder and English Art Song and

From second left: Sam, Nigel, Phoebe, Jim Bonnett, Bridget Bonnett (author's collection).

22. Nicky's Singing

again at the Cheltenham Festive and the mid–Somerset Festival. As a result of these successes, she was invited to sing at many concerts in churches and concert halls in Oxford, Newbury, Radley, Marlborough, Tavistock and Malmesbury Abbey. In these concerts she would usually be one of the four soloists, either mezzo-soprano or contralto, with a supporting choir for such pieces as Handel's *Messiah*, Fauré's *Requiem*, Mozart's *The Requiem Mass*, Haydn's *Nelson Mass*, Handel's *Saul*, Bizet's *Carmen* and Vivaldi's *Gloria*. In other concerts she would be singing a collection of classical songs such as Bach "Qui Sedes," Gluck "What is Life?," "Que Faro," Brahms "O Kuhler Vala," Mozart "Voi Che Sapete" from *Figaro*, Finzi "Who is Sylvia?" and "Fear no More the Heat of the Sun," Handel "Art Thou Troubled?," Purcell "If Music be the Food of Love" and Saint Saens "Softly Awakes my Heart" from *Samson and Delilah*.

Following are a few extracts from judges' comments at some of the festivals: (1) "Zueignung" by Richard Strauss. "A voice of splendid and special quality—nut-brown and wholesome. The song was extremely well managed." (2) "Vergebliches Standchen" by Brahms "A good understanding was shown of this song. It was conveyed through excellent German diction, and the singer and pianist rose easily to the climaxes. There was a nice change of tone colouring in the last verse." (3) "The Nightingale" by Delius and "The Watermill" by Vaughan Williams "Performance of both songs had much to commend it, lovely tone, clear diction and sensitive phrasing—a most pleasing performance and good choice of tempi." (4) "La Pesca" by Rossini. "A very creditable performance in clearly enunciated Italian, which went a considerable way to catch the mood of the song."

Of course Nicky was often asked to sing at weddings and sometimes a funeral. The popular wedding songs were Cesar Franck's "Panis Angelicus," Gounod's "Ave Maria," Handel's "Father of Heaven," Mozart's "Voi Che Sapete" and many more. I was lucky enough to attend most of Nicky's concerts. It was a huge pleasure, and it always amazed me how she would stand up alone in a strange church, abbey or concert hall filled with a huge audience and without hesitation would start

Nicky Thompson the mezzo-soprano (author's collection).

singing perfectly. She had a powerful, very beautiful tone and good range. I was always the nervous one.

We decided about 1983 to buy a flat in London because I was so often having to work late. So we bought a fourth-floor, mansion block flat in the Brompton Road just around the corner from Earls Court Tube Station. It was an Edwardian building with high ceilings and a wonderful, old lift with folding cage doors. Nicky would sometimes come up on the Thursday so that we could spend the evening in London, and she would be able to do wine deliveries to Rules Restaurant, the Arup Dining Room, plus a few others. Jack Zunz, who was the Arup Chairman at the time, enjoyed the wine, using it sometimes at lunchtime in the partner's dining room for special guests!

At Arup I was appointed by Sir Denys Lasdun[1] to join him in a competition to build a new headquarters building for IBM[2] next to the National Theatre on the South Bank alongside Waterloo Bridge. It was not surprising that we won the competition as Denys Lasdun was the designer for the National Theatre; IBM were convinced that by appointing Lasdun's team they were more likely to obtain planning on that sensitive site.

23

Air Rights Buildings (1985) and Working with Terry Farrell

Rising urban land values in the 1980s stimulated interest in the concept of air rights developments. The principal challenges of air rights buildings are engineering ones, so the design solutions tend to be engineering-driven. The key is always how do you gain access to the building in the air? With the right architect on board, the results can be innovative and often unusual building forms of distinction.

An air right is the right to use the air space above a site. A site owner can sell the air right, just as he or she has the right to sell any other part of a site. Such a sale usually leads to upper building functions that are very different from the ground level functions. There have of course always been buildings with different uses at different levels. What distinguishes the air rights building is that it is erected above an existing use which is not significantly disturbed during the construction period.

There was a sudden surge of air rights developments in and around London in the late '80s following the construction boom of the early 1980s. It was caused by the high value of land and the unavailability of large sites. This combination of circumstances forced developers and their architects to look at the possibilities of building on "bad ground." Bad ground takes many forms. It may be occupied by buildings, roads, railways, rivers or canals—or by anything else that severely limits where and how building foundations may be placed. The key to viability of an air rights development is the existence of a small area of adjacent "good ground" that will provide access to the air right. Without this good ground, development is not possible.

Existing ground level uses of the site often dictate that the superstructure has to be of unusual form and has to have long spans. Even if the special structure required is much more expensive than normal structure, its effect on the overall cost of construction is relatively small because the cost of structure in a commercial building is relatively small (it was approximately 20 percent of total construction cost in the 1980s).

One of the world's most complex and spectacular air rights developments is Embankment Place at London's Charing Cross Station (the railway station had

Embankment Place, Charing Cross, London—front elevation (Penn Station, Wikimedia Commons).

opened in 1864 on the north bank of the Thames, in central London). Here the "good ground" was not only to the side of the building but also in the Victorian vaults. This permitted the luxury of access underneath and up through a platform in the middle of the railway station. The resulting Embankment Place air rights development is also a good example of the enormously complex legal agreements required for the use of air rights. The drafting of the agreement for Embankment Place took a year to complete. It was some document. It not only covered what was to be built, including work to the station, but also covered safety for the public below—during the construction period—and contractor's permissions. The planning permission was negotiated in tandem with the drafting of the agreement. The British Rail Property Board had, of course, considerable experience in air rights construction and all the associated legal and contractual requirements. Usually it acted as a partner in the development process. Building over roads is slightly different. The highways authorities, in which control of the roads

23. Air Rights Buildings and Working with Terry Farrell

is vested, have certain rights over the road surface and the space necessary to use the road. The owners of adjacent buildings may also have rights up to the center line of the road and both above and below it. These circumstances can raise interesting if potentially formidable possibilities.

Around 1982 I was called to a meeting with Richard Burton[1] of architects Ahrends, Burton and Koralek to look into the possibility of building over Charing Cross Station and providing a travellator link over the Thames alongside Hungerford Bridge.[2] Richard Rogers[3] joined the meeting to discuss doing a similar air rights building over Waterloo Station on the south bank of the Thames, opposite Charing Cross, and connecting with the travellators to the building over Charing Cross. There was no client—this was just investigatory work such that, if we could make it work, we would search for a client. I worked out a good scheme for building over Charing Cross by having two lines of columns constructed through vaults but effectively spanning across the six main railway tracks entering the station. But the key to any development was how to enter the building, as the existing flat roof over the station was surrounded by occupied buildings. When the station was originally built in the late nineteenth century, an arch had covered all the tracks and platforms. (The arch was replaced with a flat roof early in the twentieth century.)

Later in 1982 I was interviewed by Terry Farrell,[4] who had approached John Martin, Arup's new chairman, asking John to introduce him to three of the Arup multidisciplinary groups so that he could select a group. He had been invited to do a submission for the new BBC headquarters building at the top of Regent Street (to include the building which now houses the Langham Hotel). We were delighted that Terry selected our group, and together we prepared a brilliant design and presentation. We were shortlisted to the final two but, sadly on this occasion, we lost in the final to the team led by Sir Norman Foster,[5] who also had another Arup team in support. Following this I carried out a number of competitions and presentations with Terry Farrell.

Geoffrey Wilson, Chairman of Greycoat PLC,[6] had bought up buildings along Villiers Street, off the Strand. These properties, alongside Charing Cross Station, were key to the station's redevelopment potential. From 1864 until 1906 the single-span, barrel-vault roof of Charing Cross Station had been a prominent riverside landmark. This roof had collapsed during repairs and been replaced by a lower roof of short span trusses. Now a substantial area of high-level office space might be created—premium office space close by Trafalgar Square and Whitehall. Each floor constructed above the station tracks and platforms could—with excellent planning, architecture and engineering—offer almost half a hectare (a whole acre) of office space.

The British Rail Property Board gave Greycoat one year to produce a feasible scheme and conclude a development agreement with them. Greycoat decided to commission Terry Farrell to form a proposal for building the office block over the railway station platforms. Terry asked me, and I was able to show him the work that I had previously done for building over Charing Cross. This was my own work carried out on a purely speculative basis. In the autumn of 1985, Greycoat commissioned my group to help Terry to provide the necessary engineering back up.

108 Part II—A Leader in Building Engineering and Business Development

A year seemed plenty of time. But we had not only to conceive a technical solution but also to convince the various departments at British Rail that they could accommodate the proposals. These departments included the BR engineers responsible for the existing structure, the station management responsible for the smooth running of the station and the traffic planning staff responsible for the timetables. While this was going on, Terry's people had to gain planning permission for a major office development (where none had previously existed) on a prominent site in the Thames skyline that was abutted on three sides by a conservation area. Now the real constraints began to present themselves. There were matters of logistics. For example, Charing Cross was handling 120,000 passengers per day through its six platforms—almost three times the volume of any other London main line station. The station had to remain operational throughout the project (weekend possessions were possible but had to be booked 10 weeks in advance). This fact alone explained the initial reluctance of the station management to allow any intrusion, columns included, through the existing platform environment. Planning was everything. For example, only by booking years ahead were we able to get the fabricated structural steel to site by train.

Where could we site the lift shafts to service 40,000 m² (430,556 ft²) of office space on a 100 m by 50 m (328 ft. by 164 ft.) floor plate? Ideally, they had to be central and had to pass through the centre of the platforms. We examined old records and found that a ramped cab road had risen from the old stables in the brick vaults beneath the platforms to serve the passengers at platform level. This

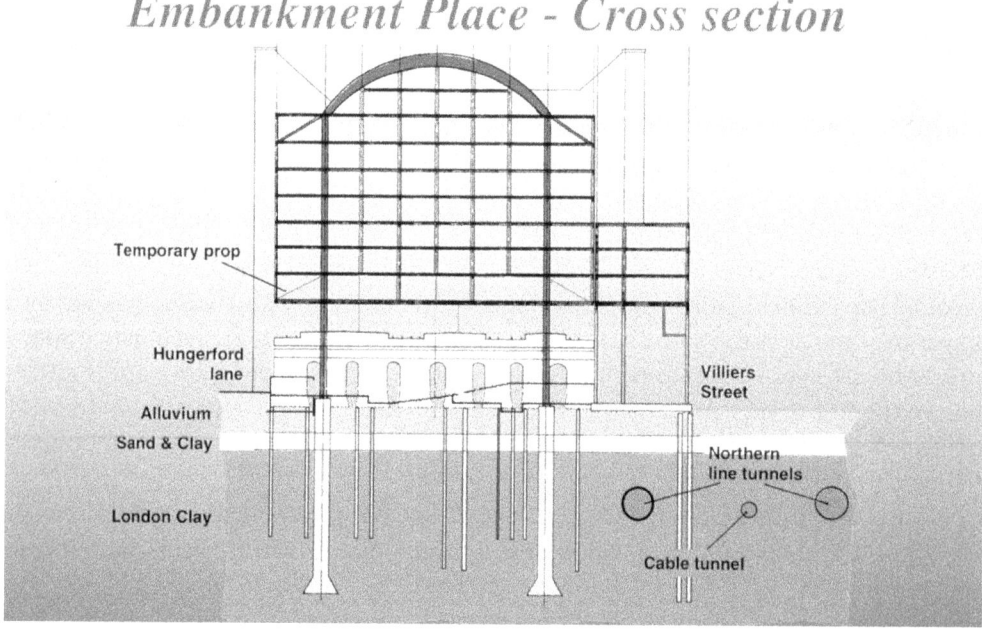

Embankment Place, Charing Cross, London, cross-section showing: Northern Line underground railway; deep under reamed piles; nine floors suspended from bowstring arch (Arup).

road had long been decked over to provide space at platform level for ancillary buildings to house station staff. But it presented an opportunity to place a hole out of the brick vaults without upsetting the integrity of the remaining structure, an obstruction at platform level that we might remove and replace with one of our own lift shafts.

The idea for lift shafts through the cab road solved the problem in the office above but presented a new design challenge. The shafts would be 70 m (230 ft.) from Villiers Street and the front door. How could the architect design an entrance hall 10 m (33 ft.) wide and 70 m (230 ft.) long with the row of columns necessary to support the superstructure down the middle? We discovered that parts of the original station structure had been founded on timber piles that had rotted, causing considerable differential settlement. The settlement had been stabilized with some underpinning which protruded below our site area, further reducing our scope for foundation solutions beneath the site. There were also abandoned railway tunnels beneath our site—but how extensive were they? Where could we place the foundations and columns?

Terry's people delved into the history of the site. The Players' Theatre had operated under the arches only since 1946 but could trace its history back to 1858 when Gatti's Music Hall started in the old Hungerford Market that had preceded the railway station on the site. The theater had to be retained. But how could we get an auditorium to work in a 10 m (33 ft.)-wide brick vault if we put a row of structural columns down the middle, as common sense dictated we should?

Now the structural design of the building began to emerge in response to the constraints of the site. The architect had been struggling to establish the urban framework for the air rights building—a strong, bold form in whitish stone, capped by a strongly modeled black rooftop. We had been struggling with the challenges of spanning the railway tracks, and it seemed that, whichever way we turned, the big span required for this would mean losing a whole office floor through meeting the need to provide a horizontal transfer structure. Then one of my colleagues, Malcolm Barrie, came up with the idea of a tied bowstring arch plus suspension structure. It was what Terry Farrell dubbed a "magical moment." The arch was as elegant an idea as was possible, and we were all instantly committed to the concept. But we still had to make it work.

The building was to be supported on as small a number of columns as possible, placed to minimize disruption to the station. The ideal solution from a planning point of view was for only two rows of columns, 36 m (118 ft.) apart. This would overcome the difficulties in the vaults by providing a huge area free of vertical obstructions and therefore suitable for the theater and entrance hall. The preferred scheme for the superstructure was worked up as a series of huge steel-tied arches built at roof level, with the office building suspended below. Plantrooms were located within the vaults wherever possible, and four service risers—like giant buttresses—were positioned outside the building, bypassing the station. These, which also housed staircases, acting with the lift cores arising up through the old side platform provided the lateral stability.

So we had a scheme combining civil and building engineering concepts. That scheme fitted in comfortably with its urban environment. It lent distinctive form to the architecture of the building. And, probably because of the logic, the various agreements were falling into place too.

Now we could design Embankment Place in detail so that someone could build it. In the overheated construction market of the time, it was necessary that we made the concept easily understandable so as not to frighten off subcontractors or attract price premiums. It was also going to be necessary for the building to be constructed at a pace comparable with other large developments being built in London to create as much competition as possible between potential tenants.

The highly visible building in the air rights over the railway platforms houses nine stories containing almost 40,000 m² (430,556 ft²) of office space. There are also nearly 20,000 m² (215,278 ft²) of additional accommodation located within the old brick arches: plant space; the theater; the impressive entrance hall; car parking; service facilities; retail units. The engineering for this hidden part of the development, while less spectacular than the offices, was just as complex in its requirements for custom-crafted details to solve the engineering challenges that are part of building in, under and around old buildings. The design development led to the building being considered in four parts. Firstly, there was the air rights building located above the platforms and suspended from the tied arch at roof level. Secondly, there were the vaults that encompass all the development below the railway tracks. Thirdly, there was the canopy building that had started out as an acoustic canopy on the front elevation but had become a two-story extension to the main office building. Fourthly, there was the Villiers Street building, a low-rise, conventional building constructed on the space between the station and Villiers Street to achieve sympathy with the streetscape and mask the enormous bulk of the building behind (internally, the Villiers Street building is part of the main office space).

The air rights building formed the major area of office accommodation and was the structure most obviously influenced by the site constraints. We had investigated and cost estimated structural solutions ranging from the traditional four-column and six-column schemes to large span two-column schemes. I mentioned earlier that the ideal was for only two rows of columns per vault to pass through the station on either side of the station, so this is what we had set out to do. We adopted column spacings of 36 m (118 ft.) across the station and 10.5 m (34.5 ft.) or 12 m (39 ft.) along the length of the station. Such an arrangement required a large transfer structure which, under normal circumstances, would be positioned at the lowest level above the station, with the building constructed on top. However, this would have required a story-height truss, resulting in a floor unsuitable for high specification office letting. Because there was a height limit on the building, this would have amounted to a lost floor. So we had investigated other methods of bridging the space—full frame, Vierendeels, portal frames, A-frames and mast and guy solutions. Malcolm Barrie's idea for an arch suspension structure led to the favored solution of hanging the building beneath a series of steel-tied arches. This provided an efficient structure and maximized usable office space. It maintained the rights of light to adjacent properties and created the distinctive barrel-vaulted appearance. While simple in concept, the scheme was inevitably complex in detail, in terms of both design and construction.

The arches spring from the eighth floor, some 40 m (130 ft.) above street level, and span 36 m (118 ft.) with a rise of 10 m (33 ft.). They consist of steel box sections 2 m (6.6 ft.) deep by 1 m (3.3 ft.) wide at the crown, and 1 m square at the springing points. The flanges of the boxes are 55 mm (2.2 in.) thick and the webs 30 mm (1.2 in.). The springing points are joined together with a post-tensioned concrete tie beam. Between the springers and the associated columns are large bridge bearings which allow sliding and rotation at one end and rotation only at the other end. Intermediate columns hang from the arches at 6 m centers. To maximize use of the site footprint, the building cantilevers 8.5 m (27.9 ft.) to each side of the main columns. We decided to support this area using a triangulated system of struts and ties at high level. Unbalanced loads between one side of the building and the other are counterbalanced by 13.5 m (44.3 ft.) long beams that cantilever beyond the main columns, in towards the center of the building and react against the dead loads of the floors supported by the arches.

The transfer structure produces very high column loads, approximately 2,800 tonnes (3,086 tons) per column and beyond the capacity of British rolled sections. The works contractor, given the choice of using a welded box section or rolled sections produced by Arbed of Luxembourg, chose the Arbed sections. These have 125 mm (4.9 in.) thick flanges and 80 mm (3.1 in.) webs and have compact dimensions. Below the first level above the station, the column sections become 1 m (3.3 ft.) diameter reinforced concrete members possessing sufficient stiffness to extend unrestrained 18 m (59 ft.) down to foundation level. Lateral restraint to the building is provided by two steel-braced cores and a large steel-plate core that houses the main lifts. We cost estimated alternative proposals for a steel-braced core, but the steel plate option proved the most economic for providing the required stiffness.

We investigated several options for floor beams. Tapered-steel plate girders were chosen for the primary beams, in place of normal rolled sections. This was an interesting choice because only recently had developments in plate fabrication for ships, offshore structures and portal frames become sufficiently flexible to produce structural elements for buildings that were both economical and in sufficient quantities for particular applications. Savings in weight more than made up for the increased fabrication costs. For a 12 m (39 ft.) span, the tapered girders are typically 750 mm (29.5 in.) deep at mid-span with a 7 mm (0.27 in.) web and 15 mm (0.59 in.) flanges.

Noise and vibration from trains and platforms beneath the building necessitated special isolation measures. The front elevation was double-glazed using glass up to 19 mm (0.74 in.) thick, separated by a 12 mm (0.47 in.) air gap; the first slab over the tracks was concrete plus insulation with an additional concrete slab on top. Despite being separated from the station structure, the main columns were subject to some vibration transmitted via the sub-soil. Analysis of the arch proved it to be an efficient filter to the forcing frequencies for all areas suspended from it, which made up a large part of the floor area. The vibration problem was restricted to areas attached to the main columns. Although the problem was considered borderline, anti-vibration bearings were introduced where practical to minimize the risk of unacceptable vibrations. In order to build the structure in a conventional sequence, we had to have additional temporary columns through the station. These were of very limited capacity. The bare steel frame was erected, together with its metal decking, to roof level. Then the arch transfer structure was built. Once the arch was in place, flatjacks at the top of the temporary columns would be adjusted so that the arches would passively pick up the load of the superstructure. Derek Pike and John Crack, who had prepared the detailed calculations, had carefully analyzed and re-analyzed all the loading variations. This was a nerve-racking moment, and John was there at midnight when the columns which had been carrying the loads of the eight floors as a temporary compression structure changed to a tension structure, hanging all the floors from the nine arches. The temporary columns between the station level and level five were then removed. The structure made a few creaks and groans but then behaved in accordance with our calculations. The deflected shape of the building was controlled by jacking points at the top of the hangers and by stressing of the post-tensioned arch tie beam. The concrete floors were then cast in a sequence that was coordinated with additional stages of stressing the arch tie beam.

Column loads were up to 2,700 tonnes (2976 tons)—so large that we had to come up with a novel approach to the foundation design. The ground conditions were typical of much of central London, consisting of fill, flood plain gravel and London Clay followed by the Woolwich and Reading Beds. Large diameter-bored piles would have been the normal solution, but the height restrictions within the existing station vaults precluded the use of a boring rig. Bill Grose, one of our bright, young geotechnical engineers, came up with the idea of using miners to hand-dig piles which would need to extend to maximum depths of 35 m (115 ft.) below ground level. We had miners come down from the north of England specially to carry out this part of the project. The piles were up to 2.5 m (8.2 ft.) diameter with an underream of up to 6.6 m (21.7 ft.) diameter. The capacity afforded by the underream enabled the shaft diameter to be kept to the minimum and enabled the depth of excavation to be kept well above the water-bearing Woolwich and Reading Beds, reducing the risk of underream "blowing" during excavation, as a result of sub-artesian pressures. I will never forget being lowered down inside the hand-dug pile to inspect the underream 35 m (115 ft.) below ground level. The first layers through the gravel were lined with concrete tubes, but once we entered the London clay—hard grey unfissured clay—there was no need for a lining. The very hard clay smelled damp, and as we were lowered inside the 25 m (82 ft.) pile shaft, the hole of daylight suddenly became very small. We reached the base of the pile, and inside the underream we were in a room of 6 m (about 20 ft.) diameter. The base was exceptionally hard, and to dig into the clay the miners had to use jack hammers.

Having looked at the air rights component of the project, let's move to the equally challenging "hidden" areas—the vaults. The 125-years-old brick vaults upon which Charing Cross railway station is built hide extensive and complicated structural works. Two stories of reinforced concrete flat slab and column construction were "shoehorned" into each vault to provide plant room areas, the theater, a wine bar, car parking, retail units and the main entrance to the building. These structures are totally separate from the existing brick vaults, both to fulfill legal requirements (that the station may be demolished at any time) and to provide vibration isolation for the new structures. The replacement of the cab road with the main lift and stability core for the air rights building was a major undertaking in its own right. It was designed and detailed to maintain stability of the existing station structure both during construction and when complete.

Within the vault areas, much fill material had to be excavated to enable construction to proceed. Very few facts were available on the nature or depth of the foundations for the existing station vaults. Because of the timing of the project, it had not been possible to carry out trial pit inspections in the time between tenants vacating the vaults and work commencing on site. Based on estimates of the loads on the foundations and on old drawings, assumptions were made regarding foundation depth and the nature of the ground. As excavation proceeded, the existing foundations began to settle excessively. My worst experience at this stage was a late-night call from deeply concerned British Rail engineers telling me that there was a sudden 50 mm (2 in.) movement of the station wall which supported the roof of Charing Cross Station. Visualizing the risk of collapse ensured I was instantly awake. This settlement occurred because a very high-existing bearing pressure (by present-day standards), coupled with unexpectedly weak ground loosened through installation of bored piles for temporary structures, resulted in an unusually high reliance on the surrounding overburden pressure for the foundations' "bearing capacity" and performance. Further consideration showed that the problem only existed temporarily. In the final condition, adequate stability of the existing footings would be preserved. We immediately took measures to ensure temporary settlement control by using a system of inclined mini (Fondedile) piles drilled through each affected foundation. British Rail trusted us to put things right, but I think even they were surprised at the speed with which we moved. Fondedile piling started the next morning. In fact, I met the client before breakfast and told him what we were doing.

The front of the building adjacent to the River Thames has a stepped appearance, and this is formed by the separate three-story canopy building. Braced frames, or shear walls, could not be used to provide lateral stability, and initial sway-frame designs proved too flexible. We achieved adequate stiffness by using a steel and reinforced concrete composite portal frame within the vault level and by using columns cantilevering up to the first level above the tracks. The columns are broken just below this level with rubber pads that provide vibration isolation for the offices above.

The final project component was referred to earlier as the Villiers Street building. Adjacent to the station, demolition of existing retail units provided space for a five-story building. The top three stories are office floors that link into the office floors area for the main building. The two lower floors comprise public walkways, retail areas and the entrance loggia. The whole building forms an impressive entrance to the air rights building behind.

Throughout construction the railway station remained fully operational. Practical completion of Embankment Place was achieved on October 2, 1990. I wonder what Samuel Pepys and Rudyard Kipling, whose houses once stood on the site, would have made of it? For me, Embankment Place looks as if it belongs—as if it has always been here and always will be here. But, looking beyond the good looks, an important element in its success was foresight. Many commercial buildings of the time were failing because they were designed to accommodate only existing technologies. Embankment Place was designed to accommodate changes in technology. The project cost £75 million (US$120.2 million at 1990 exchange rates), and the offices were let to Price Waterhouse Cooper for their headquarters. In 1997 the freehold was sold for £212 million (US$360.3 million at 1997 exchange rates) to the Sultan of Brunei.

The final word on Embankment Place is reserved for Terry Farrell who wrote:

> There were many difficulties—the steel strike occurred at a crucial time for the arch delivery, the sequencing of the tensioning of the building, and night-time deliveries when site possession time was seriously limited to keep the trains running. In the end, the building embodies much of what Nigel's team and I had discussed at the first meeting ten years ago. The result is a great engineering design, a structural tour de force that won the European Structural Steel Design Award and strong characterful architecture; each tells its own story and gets its separate credit (and criticism as is always to be expected in prominently positioned buildings); neither discipline overruns the other or attempts to steal the other's show; when the respective skills are properly combined together, the completed building is all the stronger for it.[7]

While I was working on the design for Embankment Place, I was approached by the developer MEPC to be their structural engineer for a new building to replace Lee House. The site area in London had been devastated by bombing during the war, and Lee House, a 16-story podium office block on London Wall, was part of a new comprehensive development that had been built in the 1960s by Sir Leslie Martin on Sir Hubert Bennett's 1959 plan for commercial development to the south of the new London Wall. By the 1980s office blocks built in the '60s had been found to be inflexible and poor containers for modern business.

One evening at the pub Terry Farrell and I discussed the opportunities for rebuilding Lee House. Using beer mats Terry came up with a concept of an air rights building over the junction of Wood Street and London Wall, with access gained from a new Lee House. I was so excited about this idea that I approached my client with the recommendation that they should invite Terry to give them a presentation.

The London Wall lies upon the northern boundary of the original Roman Wall around the city. Lee House is at the midpoint of the Wall. This is also the site of the old Cripplegate, which served as a point of entry and access into the Roman City of London.

In 1980 the city planners had given the owners of Lee House permission to double the size of the building. But to double its height would bring it to that

23. Air Rights Buildings and Working with Terry Farrell

of the NatWest Tower, which would be met with great opposition. Terry's proposal for a huge air rights scheme straddling the road, just as Cripplegate had, met with the clients' and the planners' approval and he was commissioned to design a new building. I asked John Pilkington to lead the Arup engineers and develop the design for Alban Gate—the new name for the project. Guy Battle (mechanical) and Chris McCarthy (structural)[8] were two bright, young engineers working on the project who helped develop an interesting, highly visible structural scheme to bridge the roads.

I had developed a strong personal relationship with Terry Farrell. He is an exceptionally talented architect and planner who was great fun to work with. Unlike with Fosters and Rogers, whose architecture often expressed the structure strongly, Terry only exposed the structure when he felt the emphasis was required. We worked successfully and enjoyably on many subsequent projects. Following on from Embankment Place and Alban Gate, we were appointed for the replacement of Moor House, another sixties block on London Wall, and then Vauxhall Cross on the Thames, generally known as the MI6 building. I also joined forces with Terry in helping Greycoat win the development for a new Exhibition and Conference Centre for Edinburgh, mostly to be carried out by our Edinburgh office. Similarly, Terry also invited Arup Hong Kong to help him win the Peak project and subsequently large developments on the Hong Kong Mass Transit.

MI6 Headquarters Building, Vauxhall Cross, London (Arup).

24

The End of the Vineyard and Arab Horse Racing (1988)

In 1988 we had to make a decision about the future of the vineyard because we had only two good years out of ten years' harvests, whilst other English vineyards had had six good years in the last ten. We had either severe frosts in late May, which knocked out the fruiting buds, and/or we had early frosts in September, preventing the sugar to rise. On the years when we had poor harvests we had had to buy in wine from another English vineyard (Three Choirs in Gloucestershire) to satisfy the market we had generated. The year of 1988 was therefore our last harvest.

In addition to all of Nicky's other activities, she used to take the dogs with her and go off miles on her Arab horse. One day when Nicky was selling wine to a couple, they observed that we had a very good looking Arab in the front field. It transpired that they trained and raced Arab horses and offered at no expense to give Count, Nicky's Arab, some basic training and see what he was like. Nicky said he is too old—13—but they replied that last year's champion was 13! Arab horses apparently matured late.

Count's first race was up at Thirsk in Yorkshire. We were not able to go as we had accepted to go to Henley with the parents of Nim's boyfriend. I asked them to telephone us—whatever the result. He had been put in a Grade III race of eight furlongs, and he stormed home, showing real class!

The next race was scheduled at Aintree[1] on the flat. We all went up to see him as it was half term. When we got there, Caroline, who was the trainer and jockey, was disappointed because the scheduled Grade II race had been cancelled and incorporated into a Grades I and II race, the last race of the day. She was disappointed as this race was with all the best horses, including the current champion. It was the most magical moment for us while we watched Count staying with the others in the middle of the field until the last furlong, when he weaved his way to the front and stayed there to win. Sadly, I was not able to see Nicky receive the cup in Aintree's famous Red Rum winning enclosure because I had to rush around a number of bookies to collect our winnings, which included many small sums for people in the village who had asked us to put bets on him. The bookies were packing up to go home as Count had won on long odds. So Nicky got the cup while I collected the winnings—a lot of money. After much champagne we had to stay the night in Liverpool!

24. The End of the Vineyard and Arab Horse Racing

Nicky holding Vayu—another winner—the horse on which Nicky would have her accident in 1992 (author's collection).

This success led to Nicky buying a brood mare at Newbury Show. Nicky always had a good eye. Samiha, after being broken in, showed she had very good speed and won on a couple of occasions before we put her to some outstanding stallions. The following year she had a mare foal followed by two stallions. Nicky and Caroline developed the Stitchcombe Arab Horse Racing Stables and at times a string of up to seven horses would leave our yard for exercise. Our horses did very well, and—because we had removed the vines and converted the vineyard back to grass—the horses did very well on any remaining vine shoots! We bought a Bedford horse truck to carry our horses off to the races.

On October 15, 1987, came the Great Storm,[2] and in London 18 people were killed. Unusually, I had stayed at Grove and not at the London flat. We received a telephone call from the police at about 6:00 a.m. and were asked if anyone was staying at our flat. They said they were asking as they were unable to open our door because a huge brick chimney had been blown down and taken most of the roof into our flat! We frantically telephoned Nim and her boyfriend. They should have been staying at the flat, but during the evening before, fortunately, they had decided at the last minute to go back to his flat as he needed a new shirt for an interview the next day. We were all extremely lucky because, if we had slept there, we would have been killed along with the 18 victims of the storm. The flat was an unbelievable mess from the brick rubble, roof tiles and timbers and years of soot. Two or three years before this terrible storm Nicky had bought an old cast iron telephone box[3] which we had taken to pieces and then shot blasted. Nicky's idea was to have it in our new flat as a shower! It was a huge struggle getting it up the stairs of the four-story mansion block because some of the sections were too heavy for the lift! Anyway, it was too complicated to make it into a shower, so we painted the box white and then bolted it to the wall in our main sitting room. After the storm the box was the only survivor; it was left proudly standing there surrounded by soot, broken roof tiles and trusses.

We rented a furnished apartment nearby for 18 months while the flat was rebuilt. Once the work was completed, unsurprisingly, we then took the telephone box back to Grove House, sold the flat and moved to a ground floor, Mansion block flat in Delaware Road, Maida Vale.

Telephone box erected and bolted to the brick walls on the fourth floor of our mansion block flat (Nicky making a call, not taking a shower) (author's collection).

25

Fiftieth Birthday and Twenty-Fifth Wedding Anniversary (1989)

In June of 1989 I reached my fiftieth birthday. Nicky, as always creative, suggested a cricket match. We put together an Arup team and a team of locals and persuaded the Puxleys of Welford Manor to let us use their private cricket pitch. We knew the Puxleys as we were neighbors when we lived at the Mill, and they used to invite some of the locals into the Manor for Christmas drinks! Nicky then hired an open-top bus which collected both teams from Stitchcombe and drove us to and from Welford for the match. It was huge fun, and I managed to score 50 runs and hastily retired not out! The only sadness to the day was that Sam was not allowed out from school for the event. I have never forgiven Wellington for that!

Sam was an excellent cricketer. One of my proudest moments was watching him playing for Wellington's first team when he was 16, scoring 93 not out and helping his team beat Eton. He was man of the match, and it was the first time he played for the first team.

I had similar proud moments with both Nim and Phoebe. With Nim it was getting into Marlborough College and doing well there. The highlight was her playing Lady Bracknell brilliantly in Oscar Wilde's play *The Importance of Being Earnest*. With Phoebe, who had difficulties at school, we were hugely proud of the progress she made with Mr. Foulds at Sibford School, Banbury, and, unexpectedly, her successes at long distance running!

In 1990 Nicky and I decided to do something special to celebrate our twenty-fifth wedding anniversary. I had recently been working in Kenya preparing designs for a new celebratory building for President Moi. During one of my visits, I was taken to lunch at the famous Muthaiga Club[1] in Nairobi with our local architect partners. It was a marvelous, absolutely classical, old colonial club. For lunch I had their traditional steak and kidney pie followed by a golden syrup sponge pudding. The walls were adorned with the heads of impala and buffalo. Another architect that we were working with had a beautiful house on the beach at Msambweni just south of Mombasa. Nim and Sam suggested that, to celebrate our twenty-fifth, we should all go on a safari. The plan was that we should all fly out directly to Mombasa and spend two nights at Nyali Beach, where I had swum

My fiftieth birthday in 1989: Nicky hired an open-top bus to take the cricketers and supporters from Stitchcombe over to Welford private cricket ground—with me in foreground wearing my pauline cricket blazer (author's collection).

during my return from Rhodesia to England in 1960, thirty years before. We would then stay at my friend's house on the beach for about five days before flying to Nairobi and staying a night at the Muthaiga Club. We would then embark on a private, tented safari.

Having decided that we would have an African safari instead of an expensive party, our children, Nim, Sam and Phoebe, said they would hold a party for us at Grove to celebrate the year of our twenty-fifth wedding anniversary, and we would hold it in April. About 60 of our friends joined us for a black-tie dinner at Grove where Nim, Phoebe and Lucy (Nicky's god-daughter's daughter, of Charles and Dides Woodward) waited on us—dressed as French maids—and Sam in black tie was the Maitre d'hotel. Before we all sat down at our carefully placed seating—both upstairs and downstairs—Sam, flanked by Nim and Phoebe, made a short speech and asked everyone to drink our health. It was a very successful party.

A week before we were due to fly to Mombasa on our holiday, Sally, my brother's wife, telephoned to say that Peter had been shot. She said she thought he would be OK as he was still talking before he went into hospital, but he had been shot in the head. I immediately flew out to Johannesburg. Sally met me at the airport. She explained that Peter and she had gone out the evening before last to a smart restaurant in the center of Jo'burg to entertain clients. As they were entering the restaurant, four Africans, who had robbed all the diners, were leaving. Unintentionally, Peter was blocking the doorway, so they shot him in the head and escaped. It took over an hour for an ambulance to appear, and then when he

25. Fiftieth Birthday and Twenty-Fifth Wedding Anniversary

arrived at the hospital it was more than two hours before he received any attention. During our drive to the hospital, Sally told me how she had spent virtually the whole night lying on the bed next to Peter talking to him while a machine breathed for him. She said that in the future he should slow down and take a less stressful job. Sally left me at the hospital to see him, and she said that Rob, her son-in-law, who had married Frances and was father to Luke, would collect me. The doctor took me to see Peter. It was a huge shock; he looked dreadful. The doctor then explained to me that there was no hope, there was nothing that they could do as Peter was clearly brain dead. He then asked me to switch off the life support system. The bullet had entered through one ear and departed the other side of his head.

I stayed in Jo'burg for another five days. Rob drove me around and helped me make all the necessary arrangements. Although Peter was to be cremated, we had an extremely nice service in a large church which was absolutely packed with Jo'burg's elite. Sally, Frances and Rob, Jude and Sarah were all brilliant—Peter would have been proud of them. I spoke at some length about Peter, our life together, Peter's life with Sally, his huge organizational ability, his three daughters, his skill at bridge and being my big brother.

Having scored 50 runs, I retired 50 not out (author's collection).

At the end of the week I flew to Mombasa in Kenya to meet the family, who would fly out to join me on our twenty-fifth wedding anniversary safari. I had a very hot and sticky night in Mombasa, grateful for the mosquito net. The following day I hired a car, met the family at the airport and drove to Nyali Beach—fantastic—a memorable holiday began. It was great to be with my family again after such an emotionally disturbing week.

After Nyali we went to stay in my friend's house. He had five servants, a chef, a houseboy, one gardener and two guards, who ironically wore builders' tin hats and carried bows and arrows. When laughing about our guards and their weapons with a neighbor we met, he replied, "Don't laugh—they are serious and killed an intruder a few months before!"

Swimming in that wonderful, warm sea was amazing. We also went deep sea fishing. Sam caught a huge marlin (about seven feet) which he had to fight for some time, and I caught a swordfish which kept leaping out of the water as it fought for freedom—my catch was about 5'6" long or seven feet if you added in the length of the sword! At a certain time of the day the beach alongside our house was covered with horrible scuttling crabs. We had Christmas day at Msambweni complete with roast turkey cooked by our excellent chef, Mohammed, and served by our house boy, Sammi.

We then flew to Nairobi, the Muthaiga Club, and, after an evening at the club, the following morning our safari guide arrived in a large Land Rover. A truck with four Africans and our camping gear had set off before us. We drove down through the Rift Valley into the Masai Mara which was teeming with game. I had never seen so many zebras, wildebeests, impalas, waterbucks and giraffes. We eventually found our team in a remote, lightly wooded spot quite near the Mara River. They had erected three double sleeping tents, each with their own tented loo and shower room alongside and a large tent with a great table in the middle. After unpacking and afternoon tea, we went out for our first evening game spotting drive. Most exciting! On returning to our camp we dined in the style of the famous film *Out of Africa* in our large tent where we were waited on, and there were lit candelabra on the table. While, outside, Masai guards had lit two roaring fires to keep the lions and hyenas away. As we enjoyed our meal, we could hear lions roaring. We spent two or three days in the Masai Mara watching lions,

From left: Nicky, Phoebe, Nigel, Nim, Sam—the Thompson family on holiday in Africa (author's collection).

25. Fiftieth Birthday and Twenty-Fifth Wedding Anniversary

elephants and hippos in the Mara River, including a hot air balloon ride game spotting.

After the Mara we went to two separate game farms in the Great Rift Valley. On both farms we went riding on horseback amongst the game. You can get really close to zebras, antelopes and giraffes on horseback as they accept horses no problem! The first farm specialized with rhinos, and our guide on foot always carried a rifle to protect us. The main danger was buffalo rather than any of the cats. At the second farm, near Lake Naivasha, where we saw great flocks of pink flamingos, we again went riding through the game where our main concern was not elephants but avoiding huge aardvark holes. The anteaters were digging for ants!

We then drove up north to the border with Somalia where the Shaba and Samburu national parks were situated. Being further north, it was a lot drier, more desert conditions. In Shaba our African team had once again gone ahead of us and erected our tents. After more game spotting, eating grandly in the evenings, and having an alarming moment fleeing from an angry elephant, the safari came to an end. I had arranged for a small, single-propeller plane, seating six, to collect us. From the plane we could see the snow on the top of Mount Kenya. To Nicky's concern I arranged for Sam to sit next to the pilot and to take control for a short period during our flight. After one last swim in Nairobi, we flew back to England after the most memorable of holidays. It was to be the last we would take as a family.

26

Kuwait City (1991)

Iraq invaded Kuwait on August 2, 1990. The United Nations called for immediate withdrawal of Iraqi troops, but Iraq's President Saddam Hussein formally annexed Kuwait as a province of Iraq on August 8. An Allied Coalition was formed to liberate Kuwait. The Gulf War began on January 16–17, 1991, with a U.S.–led Allied air offensive, Operation Desert Storm, which destroyed Iraq's air defences, communication networks, government buildings, weapons plants, oil refineries, bridges and roads. On February 24 the Allies launched a ground offensive, Operation Desert Sabre, northward from northeastern Saudi Arabia into Kuwait and southern Iraq. Arab and U.S. forces liberated Kuwait City on February 27, but the Iraqi occupying forces had looted and sacked the city.

Being an engineer who had worked on building and infrastructure refurbishment and reconstruction projects in both Europe and the Middle East, and being responsible for the development of business for Arup, I was invited to join a Department of Trade and Industry (DTI) task force charged with the responsibility of getting to Kuwait City to see what could be done to help get the city back on its feet and to land some reconstruction projects for the UK. I accepted the invitation and then worried about the logistics—I shuttled backwards and forwards across the Middle East (flying from Abu Dhabi to Bahrain, then traveling by taxi to Saudi) in order to meet up with the other task force members in Riyadh. We flew on to Kuwait in an RAF Hercules. This was propellers and earplugs stuff—a world away from the public concept of business travel.

Kuwait was black, shrouded in a dense pall of smoke from 700 burning oil wells (torched by the retreating Iraqis). The smoke cut out the light and heat from the sun even at midday, although sometimes, when the wind changed, the sun could just be seen low in the sky and glowing orange. We had to drive in cars with the headlights on all day!

Kuwait International Airport is 10 miles (16 km) south of Kuwait City. Here U.S. marines had defeated the Iraqis in a major battle that had ended the last organized Iraqi resistance in Kuwait. Consequently, the airport was a scene of destruction. Some hangars had been destroyed, one of the terminal buildings was in ruins and the control tower was badly damaged. While we were at the airport, two landmines exploded close by. It was mid–March, and I was the first British

26. Kuwait City

Nigel Thompson in front of Hercules plane at Riyadh, about to fly into Kuwait (author's collection).

Front view of Hercules on taxiway at Riyadh (author's collection).

consulting engineer to arrive in post-war Kuwait. (However—and this would be a matter of great significance—I was to find out later that 40 members of the United States Army Corps of Engineers[1] and an equivalent number of Kuwaiti engineers had driven from Saudi Arabia into Kuwait City on March 4.) My closest colleague in the DTI task force would be Bruce Boys of Laing. Bruce and I were met at the airport by the British Army and driven to the British embassy in Kuwait City. The embassy building had survived the war but had been knocked about by the SAS,[2] who had blasted down the door in case of booby traps.

Michael Weston, the British ambassador, had been the last diplomat to leave Kuwait and the first to return. I would become hugely impressed by his total commitment to serving British interests. For now, he was welcoming Bruce and me as guests of an embassy shorn of grandeur. Electricity was being supplied by a generator, and food was courtesy of the British Army Catering Corps. At noon one day I was talking with the ambassador when the generator failed, and we couldn't see each other through the "Kuwait smog." At night I slept in the piano room.

Bruce and I met the Crown Prince and government ministers to discuss reconstruction. We toured the city, assessing damage. Everywhere we were accompanied by two heavily armed "minders" from the Royal Artillery. This British Army presence enabled Kuwait roadblocks to be bypassed with ease, while grateful local people shouted out, "This is your country." Another big advantage of having "minders" was that some of the buildings we wanted to investigate hadn't been entered since the liberation. Our army escort was constantly on the lookout for booby traps, although, in the event, all they found was abandoned Iraqi ammunition, helmets and equipment.

Nigel Thompson with Royal Engineers bomb disposal unit inspecting buildings at damaged airport, Kuwait, for booby traps etc. (author's collection).

We quickly determined that there would be no big civil contracts to be won. Roads, bridges and the harbor had survived relatively unscathed. But many buildings had incurred substantial damage, particularly public sector buildings and private sector buildings with American or British connections. This damage was caused as much by looting and arson by the occupying forces as by shells and gunfire. The looting had been of two kinds. There had been "official," organized looting whereby staff from Iraqi universities had come down and removed scientific equipment, books, and computers for transfer to Iraq. High-voltage electricity pylons had been hacked down and taken away. Lifts and PABX[3] equipment had been removed from buildings. The second type of looting had been a free-for-all by the Iraqi Army over the entire period of occupation. This typically involved smashing up buildings, setting them on fire and stealing or destroying personal property.

The United States Army Corps of Engineers who'd entered Kuwait City on March 4 had begun performing damage surveys and assessments and managing emergency repairs. Within 90 days they would have restored electricity, reconnected water supplies, rendered the sanitation system partially operational, cleared roads and re-established airport operations during daylight hours. Their efforts had paved the way for American construction firm Brown & Root Vickers to be awarded contracts for damage assessments and repairs to public buildings (the bulk of the civil repair work, amounting to some US$330 million, would be completed by August 1992).

During my meetings with government ministries, I pointed out that some buildings in the city had been designed by Arup. Because the engineering drawings of such buildings were held in London, I suggested it would make sense for Arup to design the necessary remedial works. A very large building that particularly interested me in this respect was the shell-damaged and looted Kuwait Institute for Scientific Research (KISR).

Meanwhile, being on the spot and wanting to do what I could to help, I surveyed the Sheraton Hotel for its owners. The Sheraton had been shelled, vandalized and set on fire. The fire had spread easily—too easily—and the hotel had burned for 24 hours. By contrast, the city's Arup-designed Meridien Hotel had also been damaged but had benefited from a fire safety design approach which had limited the effects of fire and smoke.

My conviction was that if anyone apart from the Americans were to gain work in the reconstruction of Kuwait City, then it would be Arup. My reasoning was that Arup had all the skills required for the type of building remedial work that needed doing. This work ranged from concrete repairs—principally cleaning and resealing works—to the introduction of completely new M&E fit outs.

I'd also been able to flag up in meetings with the Kuwaiti government the Arup track record in Kuwait. This extended well beyond building engineering design. For example, the Kuwait structural plan (1970) by Colin Buchanan and Partners had been one of the first comprehensive planning studies undertaken in the Middle East. It had been reviewed in 1977 and then again in 1983. The second review was carried out by Colin Buchanan in association with Arup and the

Kuwait Engineering Bureau. A third review of the structural plan had begun in 1990 but was halted suddenly when Iraq invaded. So in 1991 Arup was party to the state's existing plan covering national construction strategy, short/long term plans for urban areas and the detailed plan for Kuwait City.

I had to return to the UK. The Kuwait authorities were positive about Arup. But the Americans had put themselves in the driving seat as far as reconstruction was concerned. I was leaving without the wide-scale engineering remit that I'd hoped for, or any major refurbishment or reconstruction project. I found out later that, before Saddam Hussein had invaded Kuwait, members of the Kuwaiti Royal Family had visited the United States and, I was told, had signed contracts with Bechtel and Brown & Root for rebuilding Kuwait in the event of an invasion. That is why, at the end, British engineers and contractors would only be able to obtain work as the "crumbs from Bechtel's table!"

However, the Saturday after my return to the UK, I was at Twickenham Stadium, the home of England rugby, as a guest of Laing for England versus France ("Le Crunch") in the Five Nations (it didn't become the Six Nations until Italy joined in the year 2000). I received a message to phone home. Nicky told me that Michael Weston had phoned out of office hours, and his call had been picked up by Arup house management's security people. I phoned the ambassador from Twickenham to be told that I had two hours to put together a proposal for the repair of KISR. I dictated a bid, and Michael Weston delivered it within the time limit. To crown an unexpectedly rewarding day, the rugby match featured one of the greatest ever tries by France wing Philippe Saint-André, and the right team (for me) triumphed, with the powerful England beating a much-fancied, pacey French side 21–19 to win the Grand Slam. Eventually, I would sign the contract for KISR. Then my colleagues David Brunt and Alan Pepper would form a second wave of consultants bound for Kuwait City.

27

Nicky's Accident

On Monday March 9, 1992, at about 6:25 a.m., Nicky drove me to Swindon station to catch the train to Paddington so that I would be in my office in good time to join the regular 9:30 a.m. Monday morning meeting of the executives of the Arup Board held in chairman Jack Zunz's office. At around 10:00 a.m. Jack's secretary called me out of the meeting to receive a telephone call from a nurse at Princess Margaret Hospital, Swindon. The nurse told me that Nicky had been in an extremely serious riding accident, had a head injury and was unconscious because of bleeding in the brain. Would I please come as soon as possible as they had decided not to operate.

I immediately took a taxi to Paddington station and then a train to Swindon. On the train I telephoned Nim, who was working at Yellow Hammer—an advertising agency. I had great difficulty in telling Nim because, from the hospital's call to me, I had understood that Nicky might be dying. Nim said she would meet me later at the hospital.

When I arrived, Nicky looked a terrible mess. She was unconscious in the emergency intensive care unit on a life support breathing machine with wires connecting her up to drips, etc. In addition to landing headfirst on the road, she had clearly been kicked in the face, had teeth missing and a cut and swollen face. The surgeon took me to a picture board which displayed two or three prints of the scans of Nicky's head. He pointed out to me the extent of the bleeding inside Nicky's brain. I asked him if he really understood what he was looking at because he had decided not to operate; he said they thought the bleeding had stopped and if they operated to remove the blood, they might well do more damage. I asked to have a second opinion from an expert neurosurgeon, Mr. John Adams from the Radcliffe at Oxford (we knew him because he had been Phoebe's surgeon). This was agreed, and the prints of the scan were immediately biked over to Oxford for examination by Mr. Adams. At this time very little was computerized, and it was not possible for Swindon to email Nicky's scans over to Oxford. We were still using telex but not yet fax!

The second opinion from Mr. Adams confirmed that there was little to be gained by an operation. We knew there would be brain damage, but we would have to wait to know the extent. Nim joined me at the hospital, shattered by how her mother looked. Apparently, Nicky had taken one of the Arabs, who needed some steady uphill trotting roadwork to build up his muscles, out on this frosty morning as soon as she had returned from dropping me off at Swindon railway station. This would have been about 7:30 a.m. She was trotting up a lane in Minal, just past the village telephone box, when a pheasant taking off or the sudden appearance of a tractor disturbed the horse so much that it threw Nicky off, and she hit her head on the road. Fortunately, she was wearing the proper jockey's helmet. Two men saw this happen. One ran down to the telephone box and called the ambulance while the other stayed with an unconscious Nicky. The ambulance came very quickly, so that Nicky was in hospital in around 40 minutes from the accident. Henry, our Labrador, carrying Nicky's helmet, wanted to go with Nicky in the ambulance. The horse and Henry made their own way back to Grove House. Nicky was in intensive care for six or seven days and was then transferred to a small ward where she remained semi-conscious and in a coma for about two months. Nim gave up her job to be at the hospital every day with Nicky and stayed every night with me at Grove. I used to drive up to London every day. At the end of the day I would then drive directly from the office in London to Swindon to sit with Nicky, have supper with Nim in the hospital canteen and then return to Grove to sleep. Sam was in Australia on a gap year, and Phoebe at school had just been selected to join the choir to sing Fauré's *Requiem* in Coventry Cathedral later in April. Phoebe got a huge shock when she saw her mother (which was not until the Friday afternoon as she was away at boarding school at Sibford). Would she recover in time to attend Phoebe's concert? Until then we knew nothing about head injuries. The doctors gently told us that it would probably be at least three months before she would come around, and then she would probably be in a wheelchair learning to speak again. The nurses told us about Headway, the charity helping patients and their families come to terms with the horrors of a head injury and the Rivermead Rehabilitation Centre in Oxford. It was at least three months before Nicky recognized us and gradually learned to speak. Walking was extremely difficult as she was paralyzed completely down her right side. Rehabilitation was a long, slow business, and she did in due course attend the Rivermead Centre in Oxford for two weeks, 18 months after her accident.

We all found it very hard initially. We hired an Australian girl, Jennifer Robertson, to keep Nicky company and drive her around. Nim went back to work, Sam went off to the University of Northumbria, Phoebe went back to school and I kept on at Arup but came back every day to cook supper. After nearly two years with Jennifer, we hired a local girl, Caroline, who had a little girl, to look after Nicky while I was off in London working. We had to get rid of the horses. One good stallion reached £18,000 (US$31,690) at Tattersalls racehorse auctioneers in Newmarket. Of the others, one we sold locally, and three quality horses were sold to a powerful Abu Dhabi family to live in luxurious, air-conditioned stables whilst the grooms, who looked after them, mostly from Yemen, lived in atrocious conditions.

The tragedy of Nicky's accident was that she lost her beautiful voice and has never been able to sing again. Her memory, both long- and short-term, was severely damaged; she also lost her senses of smell and taste, was partially paralyzed down the righthand side and had tunnel vision, which meant that she would never be able to drive again. For the last 28 years I have had to do all the cooking, so I now, out of necessity, have become a reasonably good cook!

At Arup, from 1992, the main board asked me to set up a business development unit (BDU) to help all the different groups in Arup with their own marketing.

28

Business Development (1992)

At the beginning of the 1990s interest rates in the UK were high, and the building industry was hit by recession. Construction projects were being put on hold or cancelled. Consultants who had completed preliminary project work, in the anticipation of major site works to follow, had to balance their staff numbers to match reduced workloads. Some consulting engineering firms lost their independence by selling out to U.S. companies, and others were lost completely.

Recession in the UK increased international activity in the construction industry. British consultants looked outwards for commissions. They were competing with Arup practices outside the UK at the very time that Arup's UK offices were looking overseas for work too. The business development unit (BDU) would focus the firm's intelligence on current trends, address market opportunities and increase Arup's media exposure. Communication was key. I had a very good team led by Tim Haig,[1] a young and very bright Canadian engineer with an MBA, and Ron Marsh, a confident and enthusiastic engineer, both supported by Julian Diamond, a highly intelligent IT specialist. These three could all see the big picture and do what it takes to make a difference—Ron had, for example, been "our man in Baghdad" at the time of the Iran-Iraq War.[2] Tim had a Bachelor of Engineering degree from the Royal Military College of Canada but, unlike most engineers at Arup, was commercially-orientated and, forming a strong partnership with Julian, was determined to help Arup join the twenty-first century. In addition, I was very fortunate in being able to ask Anne Kriken,[3] an American planner specializing in marketing and business development, to assist us. We identified the aims of Arup as being:

- To carry out high-quality, well-engineered designs for all aspects of the built environment throughout the world.
- To meet the real needs of our clients (i.e., help clients identify their needs).
- To provide a highly personal service.
- To be profitable.
- To provide opportunities for our staff to extend their horizons.

If we are to attain these aims, it is essential we:

- Have a strong base of work.
- Perform very well on all projects, developing strong personal relationships with our clients, architects and others.

- Gain new clients for our existing skills.
- Recover former clients, whether lost or lapsed.
- Gain new opportunities, both within the building industry and in other areas, by developing our core skills.

With the help of all parts of the firm, the business development unit attempted to identify:

- Our most promising target markets within the UK.
- Our target markets worldwide.
- The methods of communicating with these markets.
- How to develop and promote our skills in order to enable us to operate effectively within these markets.

Arup in the UK had been overdependent on buildings in the commercial sector as a source of income. This work accounted for 70 percent of total fees rendered in 1988–89, which diminished, on account of the recession having particular effect on private commercial buildings, to an estimated 56 percent in 1991–92. While the commercial buildings sector had decreased in volume, we saw an increase of 8 percent in our transportation activities and an increase in our industrial activity. We identified future prospects for the UK construction industry in 1992–93 (£1 = US$1.76) as:

- Civil (£2 billion p.a.) private/public infrastructure.
- Industrial (£5 billion p.a.) private sector—from further inward investment, particularly from Japan and U.S.A., and from strong demand for pharmaceutical plant and power stations.
- Buildings (£5 billion p.a.)—public sector relocations (e.g., Inland Revenue, Ministry of Defence), former public sector utilities (e.g., British Gas, water authorities, BT, electrical, post), public/private building associated with transport (e.g., CrossRail, Channel Tunnel Rail Link, airport development and repair and maintenance for DoT, hospitals, shopping centers and housing).

Regarding overseas markets for the UK offices, our most promising European markets were Germany, Spain and France. In Germany we had now established permanent offices in Leipzig and Dusseldorf (in addition to our existing offices in Berlin and Frankfurt am Main) to help us in this process. We were also working in Hungary and examining possible initiatives in Russia. The Middle East re-emerged, and there were considerable opportunities in southeast Asia through collaboration with our offices throughout the region. We used our American offices to provide opportunities for us to work with U.S. firms in Europe, the Middle East and Far East and used our office in Tokyo to help us work with the Japanese worldwide.

Our BDU significantly helped the whole firm, and particularly all engineers, to be more focused on maintaining and strengthening their client relationships. In addition to providing a service of the highest quality, we emphasized it was important that our clients appreciated our performance.

I became involved in a number of external committees. I was elected as the Arup representative on the Confederation of British Industries Council (CBI)[4] where I made contact with many British industrialists, bankers etc. I also represented Arup on the British Consultants Bureau (BCB).[5]

In 1992 with the President of the BCB and Prince Richard, Duke of Gloucester,[6] I led a small mission of consulting engineers to Singapore and Japan; its purpose was to promote British consultants to obtain work overseas. Being with a member of the Royal Family, doors were easily opened, and we were able to meet leaders of government and industry. In both countries the British High Commissioners and the British ambassador held parties for us where they invited all the appropriate people for us to meet. It was a huge help to us at Arup as we had small offices in both Singapore and Tokyo, and I was able to help raise our profile. Our last day in Singapore was a Saturday, which happened to be the day that the final of the six-a-side cricket international competition was to be held, and the Duke of Gloucester was invited to attend and to present the cup to the winning side. The two teams in the final were England and Australia. As the Duke did not enjoy cricket, he asked me to take his place, which I was more than happy to do, as I love cricket. Anyway, it was a most enjoyable day and quite amusing because all the cricketers were expecting the Duke of Gloucester to be there. Even though initially I kept explaining I was not His Royal Highness but his "stand-in," in the end I gave up and became royalty for the day and gave the cup to the successful English team!

It was my second visit to Japan, the first being in 1977 to hold discussions with the contractor selected to build our university in the desert in Qatar. Because we had opened an office in Japan, I was able to take the Arup engineer running our Tokyo office, John Batchelor, into the very high-level meetings with ministers, etc., which he found most useful. John Batchelor is a fluent Japanese speaker with a Japanese wife. Arup Japan had been established in 1990 and, under John's management, had quickly become the most successful foreign design firm working in that country with 16 projects already then on site. Because HRH The Duke of Gloucester was visiting Japan, the Crown Prince[7] laid on a ball especially for the occasion. Fortunately, I knew beforehand that I would also be invited, so I had taken my dinner jacket with me. The Duchess of Gloucester, Brigitte, flew out to join her husband. I had the honor of dancing with the Duchess. The Duke was originally trained as an architect, which is why he had accepted the role as President of the British Consultants Bureau. I was Chairman of the BCB for 1993–94.

After Singapore and Japan, I led BCB missions to Thailand, Malaysia, Philippines and Indonesia. With the Department of Trade and Industry I was also asked to lead construction missions to Hong Kong, China, Kazakhstan and Kyrgyzstan. The latter mission was politically led by Sir George Young, Minister for Housing, Inner Cities and Construction whilst he referred to me as his "shop steward" leading all the contractors and consultants! Both Kazakhstan and Kyrgyzstan were still effectively managed by Russians who had been sent out East to run the provinces! I had to take Sir George Young's place at some of the dinners which were laid on for us by heavy drinking Russians, because Sir George was teetotal and did not smoke—so he handed this privilege to his "shop steward." In Kazakhstan, I remember, after much vodka, our construction ministry host saying, "How in England could you be a minister if you didn't drink, smoke or enjoy some of the better things in life?" The Arup work in East Asia increased significantly, particularly due to our successful Hong Kong office.

In 1993 I became a founding member of the British Council for Offices[8] as well as being a council member of the Confederation of British Industries and a council member of the British Consultants Bureau. I'd been a member of the British Overseas Trade Board[9] between 1983 and 1986, and this experience led, in part, to the formation of the Construction Procurement Group under my chairmanship in 1993. The group comprised some of the United Kingdom's biggest contractors, property developers and consultants. Its aim was to work in partnership with construction manufacturers to increase the number of producers offering competitive products and services. The initiative was supported by the Department of the Environment which seconded a senior civil servant, Malcolm Dodds, to be director of the group. We established a task force to reduce Britain's £1.44 billion trade deficit in building materials. We took part in an exhibition entitled "Better Made in Britain" where I was able to discuss our plans with Margaret Thatcher. We launched this initiative in London on July 20, 1994, where our guest of honor was my friend Sir George Young, Minister for Housing, Inner Cities and Construction. I emphasized the importance of a strong home construction industry, from which to export: "It is inexcusable and foolhardy that as an industry we spend more on litigation than on R&D—and sometimes more on legal fees than on design fees." I explained our intention to identify product areas where there was insufficient choice from domestic manufacturers and to work with manufacturers to develop their products and services on offer.

Hong Kong and China accounted for 15 percent of Arup's work in 1992, and we had 370 staff based in Hong Kong by the end of that year. We already had a good record of working in China and saw huge new business opportunities in China's upcoming building development, privatized energy and transportation schemes. I was invited by the DTI to join Michael Heseltine, president of the Board of Trade, and Richard Needham, Trade Minister, to join a mission for business promotion in China. At the last minute[10] Lord Prior led the mission as, unfortunately, Michael Heseltine was unwell. It was successful for Arup as I had persuaded Andrew Chan, the leader in our Hong Kong office, to meet me in Beijing where I was able to introduce him to Chinese dignitaries, British officials and other international firms with ambitions of working in China. Andrew remarked that I was the first active member of the main board to visit mainland China, and he was very grateful for my help. Our first project in China had been a hotel for Gordon Wu.[11] The members of the mission were invited to a meeting in the Hall of the People in Tiananmen Square, where we met the president of China[12]. After all the meetings I took the opportunity to visit the Forbidden City. At that time the roads around Beijing were teeming with millions of bicycles which, today, have been replaced with millions of BMWs.

Airports were big business in the 1990s. They were of particular interest

Nigel Thompson with Prime Minister Margaret Thatcher and Lord Feldman, architect of the Better Made in Britain campaign (author's collection).

to me because they offered opportunities to market a big mix of mainstream and specialist engineering, project management and associated skills. This was exactly the right sort of work for Arup. We were on site with the new Kansai International Airport at Osaka, Japan, and the New Hong Kong International Airport, ground transportation centre and air cargo terminal at Chek Lap Kok.

Our work in China grew rapidly. This happened firstly through our Hong Kong office, where clients liked working with Hong Kong nationals supported by some "wide eyes" from London. In due course we had offices in China all led by local engineers. By 2020 we were the engineers for nearly 90 percent of all tall buildings in China, plus a number of power stations, railways and airports—not forgetting the "Bird's Nest" Olympic stadium!

At home, Manchester Airport was being expanded. We were on site at John F. Kennedy International Airport, New York. Planning work was underway for Lester B. Pearson International Airport, Toronto, and for Terminal 5 at Heathrow. In alliance with Aéroports de Paris, we put in a great bid for the Second Bangkok International Airport, Thailand (100 million passengers per year) but lost out to an American group when Thailand's Ministry of Transportation and Communications set up the New Bangkok International Airport Company to develop and build the airport.

28. Business Development

From 1996 to the end of 1998 I was deputy chairman of the Department of Trade and Industry's Overseas Project board's British Airports Group (BAG). This was created to provide an active and effective means of communication between government and the air transport sector to encourage a coordinated UK approach in pursuit of major work opportunities overseas. I was at the same time deputy chairman of the British Thai Business Group and was a member of the Singapore British Business Council.

My overriding business development interests were still vested in construction. We were working on tall buildings in cities including Sydney, Hong Kong, Shanghai, Manila, Seoul, Frankfurt, Berlin and London. Our wide span buildings included Brisbane Convention and Exhibition Centre, which opened in June 1995, and Melbourne Convention Centre, which opened in February 1996. Our San Nicola Stadium at Bari was a showpiece for the Italia 1990 World Cup and for its engineering team which included Peter Rice, Alistair Lenczner and Tristram Carfrae. We went on to do the City of Manchester Stadium, using sports construction for social and urban regeneration. We were involved in the Millennium Stadium at Cardiff, which had the first closing roof in the UK. We were working on the 600 ft. (183 m) span Miller Park, Wisconsin, the first fan-shaped retractable-roof ballpark in the U.S. Some of our most innovative buildings were principal visitor attractions at the World Expos in Seville (1992), Lisbon (1998) and Hanover (2000). Projects that we were working on in the UK and that would become international tourist attractions included the London Eye and Cornwall's Eden Project. In 1996 I became involved with the development work to transform the huge but derelict Battersea Power Station into a major entertainment venue incorporating shopping, leisure, accommodation and conference facilities. In 1997 I worked with Richard Burge and his colleagues at the Zoological Society of London on the proposals for the new London Aquarium.

I could write several very thick books about impressive Arup buildings of the 1990s. I've touched on buildings, and particularly the international spread of the firm's building projects, because of their contribution to our overseas earnings. But what set the 1990s apart from any other era of the firm's development was its growing success in the transport and energy sectors.

Arup work outside the United Kingdom had risen from 14 percent of turn over (1987) to 35 percent (1992). Fee income from outside the UK rose from £33 million (1991) to £35 million (1992), £51 million (1993), £78 million (1994) and £79 million (1995). Arup was named British Consultants Bureau Consultancy Firm of the Year in 1994 (for Kansai International Airport) and in 1995 (for Shajiao "C" Power Station). I applied for the Queen's Award for Export Achievement[13] for Arup in 1994 and 1995 and was successful on both occasions. The 1994 award recognized that Arup was a "consistent exporter" with overseas earnings increasing five-fold over the six years covered by the application. When the 1995 award was announced in April of that year, Arup became the first firm of consulting engineers to win in successive years. The 1995 Award again acknowledged a substantial increase in overseas earnings (to almost half the firm's total earnings of £160 million for 1993–94). It also noted that Arup had focused activities on three

Nigel Cooper Thompson Esquire receiving from Her Majesty The Queen, on her official birthday in June 1996, in the forty-fifth year of her reign, the Honor of Commander of the Most Excellent Order of the British Empire (British Ceremonial Arts Ltd.).

Nigel Thompson in discussion with Her Majesty The Queen about his work with Arup and with the Department of the Environment in improving industrial manufacture within Great Britain (British Ceremonial Arts Ltd.).

new market sectors in the Far East and Southeast Asia—power, railways and airports. In these sectors in these regions, the firm had achieved an increase in earnings from £1 million to £24 million in three years. The 1995 award was presented to Arup chairman Duncan Michael by the Queen's representative, Colonel Johnson, on November 30, 1995.

The Queen's Birthday Honours List in June 1996 included the announcement of CBE for Nigel Cooper Thompson, Commander of the Civil Division of our said Most Excellent Order of the British Empire. In October 1996 I was accompanied to Buckingham Palace by Nicky and my son Sam and daughter Phoebe. Her Majesty the Queen presented me with the award and talked with me about Arup.

Nim was unfortunately unable to join us at the palace as she was in New Zealand with her husband Jeremy Clarke whom she had married in 1995. During the reception held in our garden at Grove House, the plan was that Sam, who was the co-pilot in Tim Williams's tiger moth, would fly low over the garden and drop rose petals onto the guests. But as Tim lowered the plane, Sam undid the piece of string holding the petals in the bag when the string flew off, wrapping itself around the ignition and stalling the plane. Tim was then gliding and, fortunately, he managed to land the plane in the field behind our house while avoiding cows and overhead wires. He unwrapped the string, taking off again and flying around again, successfully dropping all those rose petals into all the open-top cars parked in the adjacent field!

The marriage, however, was not a success as Jeremy turned out to be a different person to the one Nim thought she was marrying. Nicky never trusted him, but I was taken in like everyone else, except that I did slightly question the Ferari with a JC number plate! Fortunately, they had no children, and Nim divorced him 18 months later.

In February 1998 Duncan Michael, the new chairman, wanted to reorganize at the highest level and bring onto the Main Board some younger engineers. He asked a number of us to step down from the main board. When I

Nigel Thompson standing in place of his late brother Peter to give the hand of Peter's daughter Jude in marriage (author's collection).

went home that evening, I considered my position. As Nicky had recently survived a serious horse riding accident—from which she could not fully recover—I decided that now was the time for me to take early retirement to make way for younger people at work, create more time for my family and pursue other interests. Bob Emmerson, who was now the new chairman of Ove Arup & Partners, appointed me as non-executive deputy chairman to enable me to continue with my business development.

In May 1996 my niece Jude was getting married to John Wixley, and I was delighted to be asked to represent my brother Peter and take Jude down the aisle. While I was in Jo'burg for the wedding, Nicky telephoned me to say that she had just heard from the Prime Minister that I was to be awarded a CBE for my work in assisting British business.

29

Zimbabwe (1997)

Zimbabwe in the 1980s and early 1990s was a model for African social and economic development. This was the country where I had started as a junior/apprentice between 1956 and 1960. Now for a short time I was the main board link with the leaders of the Arup practice in Zimbabwe.

Zimbabwe was a marvelous country. It could feed not only itself but also many of the adjacent countries. It has a great climate—warm during the day but cooled down at night. Zimbabwe is also a cradle of the human race. Remains of early humans, dating back 500,000 years, have been discovered here. Around 500 AD the land stretching between the Zambezi and Limpopo rivers began to be inhabited by Shona-speaking people. These people had no written language, and their traditions have not been handed down orally. But they were fantastic builders, and Great Zimbabwe, 200 miles (320 km) southeast of Harare, is one of the world's most impressive ruined cities. It covers more than three square miles (more than 800 ha) and was home to up to 40,000 people between around 1100 and 1600 AD. This is where Zimbabwe's building style—stones stacked without mortar—was developed to incorporate patterns such as chevron and herringbone, standing stone pillars (stelae), elaborate doorways and daga mouldings and plasterwork (Zimbabwe is the Shona word for "house of stone").

It is believed that Great Zimbabwe was located on a gold-rich mine. The country's natural resources include gold, nickel, copper, iron ore, vanadium, lithium, tin, platinum, chromium ore and coal. Ove Arup & Partners initially established offices in Salisbury and Bulawayo to carry out building projects in and around these cities. But Arup soon became involved in projects further afield, concerned with exploiting the country's natural resources. These included the Tokwe Dam, one of the largest civil engineering structures in Africa, and Shabani Mill, the world's most advanced industrial processing facility of its type with 350 ft. (107 m) tall main treatment plant. Even the buildings of Great Zimbabwe would have been impressed with the monolithic scale of these and other Arup constructions of the time.

Whenever Arup opened an office, the firm's policy was to create employment for local people and to transfer technology to them. Arup offices tended to survive regime changes because most of the staff would be local—they wanted to stay on both as nationals of the country concerned and as Arup people. This is what

Great Zimbabwe: The Conical Tower, a structure of unknown purpose in the Temple Enclosure. Great Zimbabwe, designated a World Heritage site by UNESCO, is the largest historic stone complex in Africa apart from the monumental architecture of ancient Egypt. Great Zimbabwe was founded in the eleventh century and building continued until it was abandoned in the fifteenth century. The edifices are believed to have been erected by the ancestral Shona. The stone city spans an area of 2.79 square miles (7.22 km2) which, at its peak, could have housed up to 18,000 people (1976, Colin Weyer).

Great Zimbabwe: The mysterious Parallel Passage, the purpose of which has evaded the researches of modern archaeologists (1976, photograph by Colin Weyer, www.rhodesia.me.uk).

happened when multi-racial elections brought Robert Mugabe, leader of the Zimbabwean African National Union (ZANU), to power in 1980 as the president of a one-party socialist state.

Arup's early commissions in Zimbabwe in the 1980s were commercial office buildings in Harare's central business district (CBD). The state telecoms company

PTC commissioned a highly serviced headquarters building, Runhare House, which takes the form of a 12-story wing and a 10-story wing separated by a central services core containing the lifts, stairs and ducts. Another big commercial office building development of the time, spanning an entire block, was the Old Mutual Centre next to Africa Unity Park. This provided 16,000 m^2 (172,000 ft^2) of open plan office space for the building's developer Old Mutual Property Investment Corporation. Runhare House was completed in 1985, and the Old Mutual Centre was completed in 1988. These and other new office buildings in Harare's CBD created state-of-the-art commercial building facilities.

Arup's success in Zimbabwe is due to the commitment of its staff in the country. They were led in the 1980s by Jeff Casson, Fred Smith and Stuart Perry. Jeff retired, and in the 1990s my principal contact was Andy Howard, a mechanical engineer with strong integrated design and project delivery experience who had started in my building engineering group at Arup and went out to Zimbabwe for experience. Andy had been based in southern Africa since 1990 and had worked on the low-energy designs for both Eastgate (offices) and Harare International School. He became chairman of the Arup Africa board in 1998 and, in 2001, left Africa to take up leadership of Arup's Los Angeles office (where he was subsequently appointed to chief operating officer for Arup's Americas region). Fred Smith and Stuart Perry stayed in Zimbabwe, providing an amazing longevity of continuous service to the firm's clients in that part of the world. After Andy's departure, they were joined as Arup leaders by Ignatius Dube and Andy Marks.

30

Financial Services Sector (1990s)

I'd foreseen that I'd need to get close to U.S. banks and their architectural consultants if I were to pick up work in the redevelopment of the London Docklands and the City of London. During the 1990s, I was fortunate to have the help of Anne Kriken, a bright, American marketing consultant who started as an architect with Skidmore Owings & Merrill (SOM) and had since worked with many different U.S. architectural practices. Anne, through her experience in New York, knew many of the U.S. banking clients. U.S. firms were much brasher than the British, who traditionally would wait to be invited to submit proposals for projects. It was through the U.S. influence that Arup and other British firms started printing brochures of their work. I remember back in the '70s when I sent a copy of our house magazine to an architect and a possible hospital client I wanted to work with—I was telephoned by Peter Dunican, who said, "Did they ask you to send them our house magazine?" As the answer was no, I was reprimanded for failing the professional etiquette of not "touting" for work. Interesting how times had changed!

Anne, as our marketing consultant, and I, as deputy chairman of Ove Arup & Partners, went on a major campaign of visiting U.S. banks and U.S. architects in London. The City of London's "Big Bang" of 1986 revolutionized the city's financial institutions and created a commercial environment which attracted participation from overseas. From across the Atlantic, the big, American institutions such as Merrill Lynch, Goldman Sachs and Morgan Stanley (accompanied by Credit Suisse, Hong Kong and Shanghai Bank and Barclays) all came to London, either to the city or Canary Wharf, to build huge new buildings with large open plan dealing floors.

Through Anne's intelligence network in New York and the experience of Arup's Iain Lyall and John Haddon in dealing floors, we pursued these banks and owners of Canary Wharf. Fortunately, I developed a good working relationship with George Iacobescu, who originally trained in Romania as a structural engineer and had come to England working for Olympia and Yorke for the construction of Canary Wharf. He was now the chief executive, and we managed to get invitations to submit proposals and enter competitions resulting in major commissions for all the banks I have just listed. The appointments varied from the complete design of a new building to major refurbishment caused by "churn." The banks were re-fitting huge dealer floors on a regular basis. They all required high-quality air-conditioning and amazing electronics.

One project above all, at that time, demonstrates the benefit of working "up close and personal" with the client and other consultants. The job was in the City of London for Merrill Lynch, which was by now a vast operation with about 60,000 employees in some 40 countries, handling client assets of $1.5 trillion. Merrill Lynch had already, in June 1996, appointed American practice Swanke Hayden Connell Architects (SHCA)[1] for the brief and site selection of a building that would concentrate its facilities on one site in London. Anne identified this project from the U.S., and I managed to obtain an interview where I was supported successfully by John Hirst, as project manager and structures, Iain Lyall, mechanical and John Haddon, electrical.

Merrill Lynch was outgrowing its existing sites at Faringdon Road and Ropemaker Place and had further increased pressure on these sites when it merged, in 1996, with the former Smith New Court Bank. The search was on for a building that would offer big floor plates to accommodate dealing floors of at least 3,700 m² (40,000 ft²) each. SHCA drew up a short list from more than 40 proposed sites. But everything was too small.

The size requirement focused attention on a post office site at Newgate which, being close to the heart of the city, presented enormous challenges to any prospective developer. Arup worked with SHCA to provide a comprehensive multidisciplinary service on the project. John Hirst led an Arup team covering archaeology, geotechnics, structural and MEP engineering, lifts and escalators, security, acoustics, audio-visual design, fire safety design, IT strategy and transport planning. We were able to benefit on all of our dealer-room projects by our in-house communications, broadcasting and IT consultancy led by Bill Southwood and Sam Shemie.

A major challenge at Newgate was that the site contained archaeological remains—including a Roman wall and medieval bastion—and was surrounded by ancient monuments and listed buildings. The Department of the Environment Planning Policy Guideline "PPG 16" (November 1990) had requested local planning authorities to make archaeological concerns a "material consideration" of a planning application. This created demand to preserve significant archaeological resource in situ by "good engineering practices." It requires sites to be evaluated archaeologically by means of a separate site investigation, which might involve geophysical prospecting and extensive hand dug trenching.

Arup had its own archaeological consultant, Richard Hughes,[2] who had worked throughout the world but was spending much time in the 1990s based in Arup Geotechnics. Richard worked with his geotechnical engineering colleagues to integrate archaeological and ground engineering requirements into the complex schemes demanded by modern urban fast-track development. Richard's activities were likely to include specifying excavation requirements, specifying contract work to approved archaeological agencies and supervising and monitoring the excavation activities.

In this case there was a key need to liaise closely with English Heritage[3] and the Museum of London,[4] which we were very experienced at doing. The result, and the reason for my attention to the archaeology, is that the remains and a display on the history of the site, are exposed to public view within the completed building. The west building provides public access to the Roman wall and medieval bastion, the main scheduled ancient monuments[5] on the site. For protection during building works, these remains were enclosed and wrapped in foam sheet under a ply lid. The "chamber" created was filled with washed sand that was vacuumed out on project completion. Subsequently, a new environment and display setting were created. The natural environment was monitored for humidity, temperature and wind movement for more than a year. The data collected was used to create conditions that would allow viewing from the foyer above and would allow visitors to walk around the remains.

It's not unusual for big office developments in the States to generate such gains in public space, but it is unusual in the United Kingdom. Broadgate showed the way forward in this respect, but Merrill Lynch set new benchmarks.

The overall height of the new headquarters building development was governed by sightlines to St. Paul's Cathedral. The main block is seven-stories on its eastern side and six-stories on its western side. The separate west building is five-stories. The architect was looking to articulate the client's vision of solidity, timelessness and stability. This led to structures being clad in stone so that the frame, rather than the glazing, dominates. The new buildings are steel framed, of largely orthodox structure based on a 12 m by 12 m (39 ft. by 39 ft.) grid. Steelwork was procured using an XTEEL 3D computer model issued electronically at tender. This was believed to be the biggest steelwork project tendered in this way.

The individual elements of the cladding, mainly stone-faced precast panels, were stacked and supported vertically at ground floor only. Because of ground constraints, principally archaeological, several support points could not have foundations. In these cases we turned a number of the steel façade frames into four-story vierendeel girders. This enabled us to support the precast units from hanging columns laterally restrained at the base. The office atrium roof is supported on 6 m (19.5 ft.) glass beams with two fail-safe devices in case of glass fracture (each beam is laminated with two sheets of glass, with each leaf capable of taking the full design load). In the event of both sheets breaking, a catenary wire comes into play to prevent collapse.

Reliability and resilience of power and cooling systems are vital to firms in the financial sector. These needs required the building to be able to operate independently of the external power network. Six diesel-driven rotary UPS/generator sets, plus seven air-cooled chillers, were craned into position at roof level. The sets are housed in a plantroom shaped to meet the building's planning constraints. To attenuate noise, the 300 tonne (295 ton) plantroom was isolated from the building frame by six anti-vibration spring supports.

All business-critical electrical systems are dual-fed from separate sub-stations, while different electrical risers serve alternate trading desks. A no-break changeover to an alternative sub-station can be brought into effect in the event of a fault. In office and trading areas, a pre-wired electrical distribution system and lighting management system were installed on a modular basis, to allow for future change.

The IT strategy was developed with Merrill Lynch's IT team. This was Arup's first major involvement in designing a building that was dominated from project outset by flat-screen technology in its IT design and trading desk layouts. Merrill Lynch is also one of the largest buildings to use its IT network for linking the control, monitoring and management of the M&E systems. The benefits of doing this extended across the installation of the BMS, lighting controls, security systems, power-monitoring, load-shedding and air-sampling. This strategy will also facilitate and help optimize work on any proposals for future changes in floor layout.

PART III

A LEADER IN RESTORATIVE DEVELOPMENT

"Kosovo is seen as one of the West's more successful interventions, not least because the military and civilian dimensions were soundly coordinated. Prior to intervention in 1999, the public sector in Kosovo—including utilities—was organized and managed by the Serb minority. These people left Kosovo, either compulsorily or voluntarily, as KFOR entered in June 1999. There was therefore a vacuum: civil government, judiciary, police, health service, telecommunications, power generation and distribution, water supply—all were in various stages of collapse. Furthermore, much of Kosovo's infrastructure had been damaged in the conflict.

This vacuum had to be filled—and quickly. KFOR did what it could in the early days, but soldiers by and large are not power engineers. The UK had acted decisively in the military field and determined to do likewise in reconstruction. The concept of a British Industry Taskforce emerged, with a rapidly developing focus on power generation: Kosovo's two old main lignite-burning power stations were on the brink of failure or worse.

Nigel Thompson, a highly experienced international engineer, was at the forefront of this British initiative. Here he recounts how the concept was successfully put into effect to the benefit of all the peoples of Kosovo."

—General Sir Mike Jackson

"Following procedures and launching tenders is the way an official operates. In principle it offers the best protection of the public financial interests. But in the case of the Kosovo power stations something more vital and more urgent was at stake: providing the country with electricity, the lifeline for light, heating, road safety, communications and even water. The welfare of the Kosovo people and therefore, the political success of the intervention, depended on the functioning of the power stations! In consultation with all concerned, including UNMIK (Joly Dixon) and the military in charge of the British sector (Maj. Gen. Dannatt) it was decided that urgent action was needed, and an emergency team of British engineers led by Nigel Thompson had to start immediately, taking over from the Royal Engineers led by Major Fuller. In a case of emergency, the speed of decision making prevailed on the respect of procedures, provided all guarantees are taken for an honest direct contracting. I had full confidence in the professionalism of the British consultants and the fairness of the contractual arrangement offered. I was sure that by taking the risk of 'interpreting' the procedures, and concluding a direct contract, I took the right decision that I was going to be backed up by my Headquarters."

—Marc Franco

31

Power in Kosovo

Pope John Paul II said (Vatican Stamp initiative—un expresso per il Kosovo, May 25, 1999): "Once again, at the closing of this century, war is raging in Europe. Once again people delude themselves into thinking their problems can be solved by turning to arms, violence, oppression; once again, as in an umpteenth tragic rerun, we are faced with grief, death, suffering, families torn apart, elderly people forced to abandon their homes, children sinking in a nightmare that will mark their lives forever, women in tears, men without a future, hundreds of thousands of exiles and refugees fleeing for safety. These pictures tear at the conscience of all ... they cannot be ignored and leave us indifferent."

Kosovo's power stations had been abandoned during the Balkans conflict of 1999. A unique public/private British industry taskforce for the reconstruction of Kosovo, working with the British military, strove to resume power generation quickly. Failure to do so would result in a humanitarian catastrophe in the Kosovan winter of 1999–2000.

Kosovo, formerly part of the Socialist Federal Republic of Yugoslavia (FRY), is geographically attached to southern Serbia and is in the middle of the Balkan Peninsula. Its population in the 1980s was approximately 92 percent Kosovar Albanians and 8 percent Kosovar Serbs. The province had a president, prime minister, parliament and all the ministries except for the Ministry of Defence, which was common for the FRY.

In 1989 Slobodan Milosevic became president of Serbia. He altered the status of Kosovo, removing its autonomy and bringing it under the direct rule of Belgrade, the capital of Serbia. A process of introducing Serbs into the positions of authority in the workplaces of Kosovo began. The Kosovar Albanians became second-class citizens. A "cold war" developed. Tensions led to hostilities breaking out between the FRY authorities and an emerging Kosovo Liberation army (UCK) through 1996 and 1997. The UCK came out in its military uniforms for the first time during the funeral of the Albanian teacher Halit Geci, who had been killed by Serb police in November 1997. This led to open conflict between Serbian military and police forces and Kosovar Albanian forces. During 1998 more than 1,500 Kosovar Albanians were killed, and around 400,000 people (out of a population of 2,300,000) were driven from their homes.

The international community was gravely concerned about the escalating conflict. Under diplomatic pressure, backed by the threat of NATO air strikes,

Displaced Kosovar refugees follow the railway lines to avoid landmines (UN Photo by R. LeMoyne, March 1, 1999).

President Milosevic agreed to withdraw forces from Kosovo, cooperate in bringing an end to the violence and facilitate the return of refugees to their homes. In 1998 KVM (The Kosovo Verification Mission) was an OSCE Mission (the Organisation for Security and Co-operation in Europe) set up to verify that the Serbian and Yugoslav forces were complying with the UN October agreement to end atrocities in Kosovo, withdraw armed forces and abide by a ceasefire. Part of the KVM team arrived in Pristina a month after the October 16 agreement.

Immediately after the agreement neither side adhered to the ceasefire, state loyalist forces continued to shoot at civilians and there were sporadic KLA attacks on state forces. When a KVM team arrived at the Račak massacre, they found "36 bodies, 23 of which were lying in a ditch with gun shot wounds." The reporting of this massacre led to worldwide alarm. Peace negotiations held in Rambouillet, near Paris, from February 6–23, and in Paris from March 15–18, led to the Kosovar Albanian delegates signing the proposed peace agreement. But the talks broke down without a signature from the Serbian delegates.

In 1999, there was an increase in ceasefire violations. There were acts of provocation on both sides, and excessive force was used by the Serbian Army and special police. As risks increased, the KVM team withdrew to Macedonia and immediately, state forces began a campaign of killings, rapes, detentions and deportations of the Kosovar population.

Serbian military and police forces stepped up the intensity of their operations against the ethnic Albanians in Kosovo, and tens of thousands more people

31. Power in Kosovo

Former Yugoslavia: My roles in the Balkans covered firstly Kosovo, then Serbia and Montenegro, and then South Eastern Europe in general. Map shows Kosovo to 1992 (CIA).

fled their homes in the face of the offensive. U.S. Ambassador Holbrooke[1] flew to Belgrade to persuade President Milosevic to stop attacks on the Kosovar Albanians or face imminent NATO air strikes. Milosevic refused to comply, and, on March 23, NATO commenced air strikes.

On June 10, 1999, after an air campaign lasting 77 days, NATO Secretary General Javier Solana[2] announced that he had instructed General Wesley Clark,[3] Supreme Allied Commander Europe, to temporarily suspend NATO's air operations against Yugoslavia because the full withdrawal of Yugoslav forces from Kosovo had begun. Also on June 10, the UN Security Council passed a resolution, UNSCR 1244, welcoming the acceptance by the Federal Republic of Yugoslavia of the principles on a political solution to the Kosovo crisis, including an immediate end to violence and a rapid withdrawal of its military, police and paramilitary forces. The Resolution—adopted by a vote of 14 in favor, none against and one abstention—announced the security council's decision to deploy international civil and security presences in Kosovo, under United Nations auspices. These presences were to comprise an international security force (KFOR) and a UN-managed civil administration, the UN Interim Administration in Kosovo (UNMIK).

At the end of May I had received a phone call from Colin Adams, executive director of the British Consultants Bureau (BCB). Colin knew all about my time as chairman of the BCB, my mission to the war-damaged Kuwait City in 1991 and my experience in promoting UKplc with DTI and its working groups. He knew that I was semi-detached from Arup, going into the office just two days a week, and hoped that I might be persuaded to juggle my commitments to take on an extraordinary challenge. Colin wanted to lobby the British government to establish a public/private British industry taskforce for the reconstruction of Kosovo, and he wanted to nominate me to lead that taskforce. If I could make a difference to the situation in Kosovo, then I very much wanted to help, so I said "yes."

Colin put his idea to Sir David Wright, chief executive of the recently formed British Trade International (a new body comprising an amalgam of the Foreign & Commonwealth Office and Department of Trade and Industry). Sir David liked the idea and referred it to Stephen Byers, the then secretary of state for trade & industry. Stephen Byers brought it to the prime minister's attention. Tony Blair was enthusiastic and, on June 10, in his Downing Street address on the Kosovo peace deal, spoke of helping the Kosovo people rebuild their shattered lives and communities.

That same day I was called down to DTI to meet with Keith Levinson, David Arathoon and Grahame Atkinson, who said they'd been asked by Sir David Wright to set up the taskforce. They invited me to lead it and to decide who should be invited to join the taskforce. Members would include representatives of government, of the DTI public/private partnership working groups, the BCB, Crown Agents and CBI. Having someone from the private sector to select the individuals to serve on the taskforce saved time and heartache. Civil servants—when receiving complaints from other parties about why they weren't on the taskforce—could shrug their shoulders and say it was Nigel Thompson who made the selection.

The first meeting of the taskforce was held under the chairmanship of Industry Minister John Battle at the DTI on June 15. John said that the current political vacuum in Kosovo presented a unique and challenging set of circumstances that no government had had to deal with before. He flagged up the need for lateral thinking and innovative actions—supply chains had to be built, and there was a need to deal with utility supplies still (as he spoke) under the control of the Serbs. We identified our principal objectives as being: to help the Kosovo people rebuild their province and its economy; to see how we could help British firms do the work in a joined-up way with the British government; and to work effectively with the United Nations, European Union and other international funding institutions. John Battle said that Stephen Byers wanted to get the taskforce out to Kosovo at the earliest possible opportunity; to assess what needed to be done and the potential scope for UK businesses to assist with the regeneration of the province. Assuming a smooth withdrawal of Serbian forces and knowing that the Ministry of Defence was aware of the importance of getting us into Kosovo, a mission during the week commencing June 28 seemed realistic.

Meanwhile, on June 24, I visited Brussels with Mark Gibson, head of Market

Sector Directorate at British Trade International (the DTI and FCO trade promotion body). On June 12, UN Secretary-General Kofi Annan had presented to the UN Security Council an operational concept of the United Nations Interim Administration Mission in Kosovo (UNMIK). In visiting Brussels, Mark and I were seeking to learn more about UNMIK and to meet key players in UNMIK and the EU, which would be the funnel through which reconstruction funds would move.

In 1998 Sergio de Mello was the UN Undersecretary-General for Humanitarian Affairs and Emergency Relief Coordinator. He was made a special UN envoy in Kosovo in 1999 for an initial period of two months after the UN took control over the Serbian province. He held this position until the UN appointed Dr. Bernard Kouchner. I was fortunate in meeting Mr. de Mello when I first visited Pristina, and he was very supportive of our mission. I understand he had developed a strong working relationship with Lt. General Sir Mike Jackson.

UNMIK was being formed as a single operation, under Dr. Bernard Kouchner,[4] with four "Pillars": Pillar I, Humanitarian Assistance, was being led by the Office of the UN High Commissioner for Refugees; Pillar II, Civil Administration, was directly under the UN; Pillar III, Democratisation and Institution Building, was being led by the Organisation for Security and Cooperation in Europe (OSCE); Pillar IV, Economic Reconstruction, Recovery and Development, came under the EU agency.

Among the key people that we met in Brussels were Joly Dixon, who was subsequently appointed Head of Pillar IV; Marc Franco,[5] Head of the EU task force that was to be established in Pristina, Kosovo and Marc's deputy, Therese Sobieski, who would also be based in Pristina. UNMIK and its four pillars would act as the government, and they, through Pillar IV, would request the EU task force to implement reconstruction. The EU task force would provide the funds and would use EU procedures to implement the work. All our meetings confirmed our conviction that we needed to get into Kosovo quickly to see the damage, establish what we could do to help, identify sources of funding and carry out project work that would contribute towards getting the province back on its feet.

Back in the UK, an exploratory team from the task force was mobilized. The members, their firms and their fields of expertise were: John Battle, industry minister; Nigel Thompson, leader; Tim Askew, WS Atkins—water; Mike Viney; Mott MacDonald—power; Bruce Russell; Taylor Woodrow—infrastructure and general contractor; John Anderson, managing director of Bovis Construction; Don Cook, Crown Agents—logistics, transport and supply chain issues; Stephen Wordsworth—Foreign & Commonwealth Office regional assistance team.

We took off from RAF Northolt, on the Queen's Flight, on Monday June 28. On landing at Skopje, we were picked up by Chinook helicopter and flown by an American "Rambo" pilot with a marine machine gunner at an open hatch overlooking burning houses as we weaved our way over the mountains to Pristina. This was the first business mission to enter Kosovo since the cessation of hostilities and, as such, we descended into a spectacular landscape of raging fires, burned-out housing and bomb-damaged infrastructure.

Lt. General Sir Michael Jackson,[6] Head of KFOR, welcomed us to Kosovo and explained what they had been doing and the current status. I'd been looking forward to meeting General Jackson because of a remarkable incident on June 12, when he'd narrowly avoided a potentially disastrous situation. What had happened was that a column of about 30 Russian armored vehicles carrying 250 Russian troops had headed for Pristina International Airport ahead of the arrival of NATO troops. This action jeopardized the whole peacekeeping operation. General Jackson—who speaks Russian—flew by helicopter to meet General Viktor Zavarzin,[7] commander of the small Russian force. Sheltering from heavy rain in the wrecked airport terminal, Jackson shared a flask of whisky with Zavarzin, leading to a warming of relations. General Jackson further endeared himself to the Russians by providing meals for them (they had exhausted their rations and had no supply chain). General Jackson's own son Mark—also a Russian speaker—led a squad of British soldiers assigned to protect the Russians. That evening General Wesley Clark was concerned with the possibility of more Russian troops being flown in, even though NATO controlled the airspace (Russia had placed several airbases on standby and had prepared battalions of paratroopers to depart for Pristina on military transport aircraft). Fearing that Russian aircraft were heading for the airport, Clark planned to order helicopters to block the runway and requested helicopter support.

Boarding Chinook again, with Minister of State for the Department of Trade and Industry John Battle (in foreground, facing away from camera) (author's collection).

The following morning, Sunday, June 13, Clark arrived at Jackson's HQ in Skopje. It was pointed out to Clark that the Russians were isolated and could not be reinforced by air and that Russian support had been a vital part of getting a peace agreement; antagonizing them would only be counterproductive. Clark refused to accept this and continued to order that the runway be blocked, claiming to be supported by the NATO Secretary-General. Jackson refused to enforce Clark's orders, reportedly telling him "I'm not going to start the Third World War for you." When again directly ordered to block the runway, Jackson suggested that British tanks and armored cars would be more suitable, in the knowledge that this would almost certainly be vetoed by the British government. Clark agreed.

Jackson was ready to resign rather than follow Clark's order. The British Ministry of Defence authorized British force commander Richard Dannatt[8] to use four armoured brigade to isolate the airfield but not to block the runways. Clark's orders were not carried out, and the United States instead placed political pressure on neighboring states to ban Russia from using their airspace to ferry in reinforcements. Russia was forced to call off the reinforcements after Bulgaria, Hungary and Romania refused requests by Russia to use their airspace.

Nigel Thompson flying Chinook! (author's collection).

Negotiations were conducted throughout the stand-off, during which Russia insisted that its troops would only be answerable to Russian commanders and that it retain an exclusive zone for its own peacekeepers. NATO refused these concessions, predicting that it would lead to the partition of Kosovo into an Albanian south and a Serbian north. Both sides eventually agreed that Russian peacekeepers would deploy throughout Kosovo, but independently of NATO.[9]

Kosovo had been divided into five zones essentially run by the French (North), British (Centre), Americans (East), Germans (South) and Italians (West). What came to fascinate me about the zones was how they conjured up a caricature of the different ways in which the world works. The British were terrific—they'd had plenty of peacekeeping experience and, from private soldiers to generals, knew how to diffuse tension. Their readiness to mix with the locals, famously wearing berets rather than combat helmets, relaxed the people of Kosovo. While I was there, a civilian was shot dead for failing to drop a weapon, but this extreme event received full backing from the locals, who no more wanted gangsters in their midst than KFOR did. In the USA sector the Americans were very jumpy. They were under instructions not to leave their armored cars and had to wear full body armor, which was extremely uncomfortable in the heat of June. Basically, the American authorities were petrified of having to send out body bags. The Italians were particularly laid back, preferring to sip coffee in the bars while the Kosovo Liberation Army gave a credible impression of being in charge. The Germans, being mostly conscripts, were terrified they would have to shoot someone. And the French didn't let us into their sector, saying that it was French territory.

The United Nations was to receive criticism in Kosovo from both Kosovar Albanians and Kosovar Serbs and from the European Union and DFID. I was tremendously impressed with the people at the top of the UN, many of whom were working up to 23 hours a day. Lower down, the quality was more variable. The sort of thing that irritated people was the fact that a maintenance engineer was unable to get transport out to fix a coal conveyor, while the UN car park was full of gleaming, white 4×4s sitting waiting for their users to complete their day in the office.

Back to that initial briefing, I'd requested—via the MoD prior to meeting General Jackson—that I wanted the general to help us identify a "flagship" project that we would establish as our British contribution to reconstruction in Kosovo. General Mike said, "You will have to talk to Major General Richard Dannatt who is in charge of the British Zone."

But, as were leaving, General Jackson took me aside and said, "Go and look at the power stations. They are in the British zone." The power stations were already on the schedule, but that comment confirmed my own line of thinking and heightened my existing interest. Apart from meeting the energy needs of Kosovo, the power stations supplied the power that was necessary to pump the province's water and run the province's hospitals, general lighting and some heating. The hospitals had been looted of all their equipment and computers—everything had been taken. The buildings were ransacked even to the extent that body parts from the morgues were being carried around in the mouths of dogs in the hospital grounds.

We visited the power stations. Kosovo A had been built in the 1960s and

Kosovo A Power Station (Brian Stone).

31. Power in Kosovo

added to in the '70s. It was shut down and in a state of gross neglect. Kosovo B, built in 1983, was in better condition but was also shut down. Before the conflict these power stations had also been supplying other parts of Yugoslavia—Serbia, Macedonia and Albania. The lignite fuel, a dirty, brown coal in a state between peat and bituminous coal (and a serious health hazard), had been supplied from opencast mines by a 1 km (0.6 mile) conveyor that was barely in operable condition. The Serbs had stopped Kosovo Albanians from getting into management roles throughout Kosovo, and the power stations were no exception. But when KFOR came in, the Serbs left. The mines had also been abandoned, the diggers removed and the mining plant partly vandalized. Major Joe Fuller of the Royal Engineers was the military man who had been put in charge of the power stations. He had a degree in agricultural studies from Reading University but, more pertinently, had a brother in New Zealand who had previously worked in a power station in the UK. So there was some critical messaging going on over the airwaves. The power stations were in truly appalling condition, having received no maintenance for 10–20 years. They were existing in a dangerous state and were like

From left: (foreground) Mike Viney (Mott MacDonald) opposite John Battle (Minister of State for the Department of Trade and Industry), Major General Richard Dannatt (Commander of British Forces in Kosovo), Nigel Thompson (Team Leader, British Industry Taskforce for the Reconstruction of Kosovo) (author's collection).

a ticking time bomb, liable to blow right then or at any time. Being in the power stations was more dangerous than walking around outside among the unexploded bombs. In this fraught situation Joe Fuller was trying to create an environment in which both Serbs and ethnic Albanian workers could return to work. He expressed the need to get a management team in place as soon as possible so that he could deploy his troops.

Despite the huge challenges in Kosovo, it was time for the task force to return to the UK. On the way out, through Macedonia, we passed the EU task force going into Kosovo, proving how fast off the mark we'd been. Just a couple of days later, on July 1, John Battle held a conference in London at which we fed back the findings of the task force members to some 250 representatives of British industry.

I now had meetings with DTI and the Department for International Development (DFID). On June 29 the secretary of state for international development, Clare Short, had written to Alan Milburn, chief secretary to the treasury. The memo, which had been copied to Tony Blair and to appropriate government departments, stated the need for all assistance to Kosovo to be based on sound economic principles and stressed the need to control costs and provide value for money. Bilateral aid must be proportionate, pay due regard to burden-sharing and focus on adding real value. Temptations to make political gestures or set up significant independent operations were to be avoided. Contributions should be made to the UN trust fund—energies should be concentrated on ensuring effectiveness of the international effort. These criteria provided a framework within which my forthcoming decisions on the ground in Kosovo would be made. Before that, there were urgent issues to be addressed in the UK. DFID has people engaged in engineering work and people of engineer-in-charge capability. DFID also hires individuals. Through the DTI working groups I collected CVs of people with the appropriate skills to help make an engineering team to deal with airport, hospital and water issues in Kosovo. But my bigger ambition was to put together a power engineering team to deal with mining, generation, transmission, distribution and restructuring the whole operation. To carry out this work, I attracted volunteers from PowerGen, National Grid and Mott MacDonald. But DFID considered the power challenge in Kosovo too big an issue to deal with and declined to fund this initiative.

I had to lower my sights and find an alternative source of funding. So I got back to Keith Levinson of BTI, with the proposal that a team of six specialist power engineers do one month's work, under the auspices of the DTI Power Sector Working Group, in re-establishing a secure power supply for Kosovo for the coming winter. Using DFID rates, the sum total for this task amounted to £98,000 (US$158,578 at 1999 exchange rates), assuming that local transport, accommodation and subsistence (food and water, etc., military escorts and other reasonable support) would be provided by the military. This British Trade International Special Engineering Team would work in collaboration with the DFID Infrastructure Engineering Unit in Pristina, with KFOR and with the EU Task Force. Keith accepted the proposal on the same day that I made it.

My overriding concern was that any time lapses in things happening on the power station front would result in a situation in which thousands of Kosovars could die of cold in the winter months. Now at least I had a month's work in place. The team was led by Mott MacDonald's Brian Stone and included specialists from PowerGen, National Grid, Scottish & Southern Electricity and Taylor Woodrow for mining the lignite. Its priority was to get Kosovo A and Kosovo B power stations up and running and to identify repairs required to the transmission system. The team would carry out a system survey to identify problems and would put in place procedures and contingency plans to see Kosovo's future development. On August 16 I flew out to Kosovo to check on progress of the UK power team, accompanied by Mark Watchorn of DTI, Mike Viney, Colin Adams and Professor Peter Street, who is an agricultural expert and director of Genus, the international consultancy division of Hunting Engineering. Also travelling with us was Mike Gordon, a journalist with *Contract Journal*. We were to visit Kosovo A on August 17. My wish was to find a way of keeping the power team in place despite the news from the EU that power sector contracts were to go out to competitive tender.

It was good to meet up with Brian Stone again at Kosovo A. Brian's opinion of the Kosovo power stations was pretty similar to mine. He thought that if we were to run a power station in western Europe in the condition of Kosovo A and Kosovo B, then we'd be thrown in jail. He had tales of boilers leaking and soot and ash penetrating the boilerhouse. Particularly off-putting was the flashing arc that you got when you flipped the switch gear and the red, glowing cracks in the boilers—horrific! Brian was convinced that something would blow up or that someone would choke to death—he just didn't know which would happen first.

Next stop was UNMIK Headquarters in Pristina, where I knew I must do everything I could to keep the power team on the ground. Alan Baron of TAFKO confirmed that any further contract would have to go out to tender to five preferred bidders. I said this route would take months, giving no chance of providing desperately needed power supplies to the local people for the winter. My argument cut no ice. I then met with Professor Hilary Lee, UNMIK's minister for energy, who expressed grave concern over the condition and stability of the power stations. But he said that his hands were tied by the EU procurement policy. The one encouraging piece of information to come out of the visit was that Joly Dixon, head of Pillar IV, was convinced by my argument that the power team should be kept on if at all possible, otherwise people would die and he would be responsible. The only hope left was a meeting with Marc Franco, scheduled for the next day, August 18. I didn't want this to be a big meeting but did want to have Mike Viney, Mark Watchorn and Brian Stone with me. Marc Franco was hugely appreciative of the work that the UK power team had put in and he was extremely impressed by the British Army. He wanted our team to stay but said that it would be necessary to obtain the lowest cost by applying the mandatory EU tendering procedure to first prepare reports on the state of the power stations and then go out for a second round of bidding to obtain the contractor to restore the stations.

I said that playing strictly by the rules would mean nothing would happen for months, whereas my team, already on site in the wake of experience gleaned by the Royal Engineers, could instantly take control of the plant. It was a choice, as we the task force saw it, between the rulebook and a race against disaster. Mike Viney added that Kosovo A was threatening to blow up at any minute. I said that if we were to start work now, it would take between six to eight weeks to get the damaged Kosovo B up and running. But if the job were to go out to tender, then we would be right into the winter with no power. I said that would mean letting down not only the local people but also the Serbs. That was because about 85 percent of Serbia's fuel tanks were damaged, and they had been used to heat whole districts. So the Serbs would now need electricity in the winter and would be looking to Kosovo for the supply. I said that I wasn't just "batting for Britain"—I was battling for lives.

Of course, Marc Franco was fully aware of all these arguments. But his reaction still came as a surprise to me. Against the prevailing ethos in Brussels, and against the views of most colleagues, he proposed to fund 50 percent of the management contract if BTI were prepared to match that contribution. I agreed on the spot because, apart from the task force being committed to the cause of power for Kosovo, the proposal met the criteria on aid for Kosovo that Clare Short had outlined to the Treasury on June 29.

Marc Franco said that, as we were agreed, he was prepared to withdraw the tender letters. He said that he would need to consult with UNMIK colleagues on the terms of reference to see if this was acceptable to them but indicated that he didn't anticipate problems. A management contract was drawn up that afternoon for Mott MacDonald, acting on behalf of the DTI Power Sector Working Group, to run the Kosovo power sector (including coal) over the winter under the supervision of UNMIK and a joint works management committee. TAFKO would pay for the consumables for the power sector and for workers' salaries. Bigger items would have to be approved by the joint management committee before going to tender. The management team would need to follow EU procurement rules but would do so under new, flexible procedures, given the emergency nature of their work. I knew that I was successful in my proposition when I was telephoned by the French minister for energy suggesting that we should form a joint consortium for the work. I thanked him but politely turned him down saying we already had the team in place carrying out the work!

There was, of course, some unhappiness among those who had wanted to bid for the power contract and hadn't had the opportunity to do so. European Commissioner Chris Patten helped smooth some ruffled feathers and eliminate some lingering impediments. But, essentially, the contract went ahead because it was clearly the right thing to do for the Kosovan people.

The six-man team was fully operational by September 28, managing 6,000 people, when—just weeks earlier—the power stations and mines had been abandoned. In this respect it was impossible to praise the British military—BRITFOR—too highly. They earned universal acclaim for the role they played in helping restore the province's utilities, firstly on an emergency basis and then in collaboration with the British Trade International power team and DFID specialists in the water sector.

31. Power in Kosovo

Brian Stone, who was excellent in his leadership of our task force power team, made the following comments in his lectures on Kosovo:

> There was absolutely no PPE for any of the power sector workers. The only "hard hats" were for men working in the boilerhouse, which were made from folded newspapers in a vain attempt to keep the dust out of their eyes, ears and hair. Dust in the boilerhouse was a major hazard to workers, and, over the years, it had accumulated to such a depth on top of the boilers that we were concerned at the possible failure of the boiler support hangers due to this additional load, which would have had devastating consequences.
>
> After one boiler had tripped out of operation, the temperature dropped so rapidly (it was -20 deg. C outside) that the coal froze together at the bottom of the boiler and blocked the coal feed before we could re-start it. When a workman tried to dislodge the frozen lump in order to start up the boiler again, it fell on him and crushed him. Sadly we could not get him out quickly enough to save his life. We were also aware of another workman who was killed when trying to erect an overhead line tower in an unorthodox manner, due to a shortage of the right tools and inadequate working methods.
>
> There was a constant tension between Albanian and Serbian Kosovars despite our best attempts to bring them together. Shortly before the power team arrived in Pristina, KFOR had set up 5 committees comprising Serbian and Albanian personnel to run each of the five departments of KEK (i.e., coal mines; power stations; transmission; distribution and corporate governance). We held a regular weekly meeting all together in one large room when discussions had to be held in Serbian, Albanian and English. This was a desperate attempt to bring both ethnic parties together to jointly manage the whole power sector. However, within the month of August 1999, the Serbian contingent had become so terrified that they fled their posts and were not seen again.
>
> When we arrived for our first meeting of KEK and EU TAFKO at the end of July, we found a pool of dried blood on the pavement outside the KEK office entrance. This was the result of a Serbian staff member being shot dead on the previous day when he turned up for work. We proceeded with our meeting under armed guard courtesy of the British Army.
>
> Our first attempt to meet the Serbian manager of A Station at the power plant was aborted when he was turned away at the Station gates and then shot at as he quickly drove away. He telephoned later from Serbia to apologize for missing the meeting for reasons beyond his control.
>
> Without the efforts of such unsung heroes, who worked and lived under terrible conditions, all of our endeavours would have come to nought.

Marc Franco's courageous decision to set up the management contract was fully vindicated by events because demand for power increased substantially even before the onset of winter. As of the end of August 1999, some 90 percent of Kosovo Albanians who had fled the province in early 1999 were estimated to have returned—an unprecedented pace of return that demonstrated strong commitment to restoring normal life. That normality was facilitated by the EU and DFID providing toolkits for locals to rebuild housing themselves—an aspect that made me think of Kosovo reconstruction being more B&Q[10] than Brown and Root. This all heightened the importance of securing the power supplies.

On September 20 I had cause to respond to a letter in the *Times*, and my reply included the points that, "Unlike in Kuwait and more recently Bosnia, the task force has shown how effective and efficient a partnership between the private sector and government departments working together can be. So much so that the European Commission has had complaints both in Pristina and Brussels about the British being commercially aggressive and behaving more like the French."

There was another visit to Kosovo. On October 13, I accompanied Minister for Trade, Richard Caborn, and a party including Andrew Morling, David

UNITED NATIONS
United Nations Interim
Administration Mission
in Kosovo

NATIONS UNIES
Mission d'Administration
Intérimaire des Nations Unies
au Kosovo

21 March 2000

Dear Sir,

Subject: Kosovo Power Sector Management Services Project

The purpose of this letter is to thank you on behalf of the United Nations in Kosovo for the splendid work that your Project Team has done to maximize the electrical power supply in Kosovo, following the recent conflict in the territory last year. The Team's work last summer was funded by British Trade International, who continued to provide co-financing for the management services project that followed, for which we are most grateful.

My staff and I have been impressed with the truly dedicated and professional way in which the Team has carried out their work, often under most trying circumstances. The power stations in particular have suffered from years of neglect, and I have no doubt that without the perseverance of the members of your Team in advising and encouraging a largely inexperienced workforce, the shortages of power would have been far worse than they were this winter. Furthermore, I am sure that the task of the new team, which will take over the responsibility for managing the power sector in April, has been made easier by the many improvements initiated by your Project Team.

I would be most grateful if you could pass on my sincere thanks to each of them, on behalf of the United Nations in Kosovo.

Yours truly,

Bernard Kouchner
Special Representative of the Secretary-General

The work of the British Industry Task Force for the Reconstruction of Kosovo was universally acclaimed to have been of huge success and benefit in restoring power and saving lives (courtesy Brian Stone).

Arathoon, Mike Viney, Alasdair Frew and Laurence Biglow-Smith. Richard Caborn's abiding memory of Kosovo was the noxious, yellow fog spewing from the lignite-burning power stations. That image would live in my mind too. But we had succeeded in providing the power.

There are a couple of footnotes. On June 2, 2000, Richard Caborn was able to say that, "We are very well respected in Kosovo because of the lead the Prime Minister gave to the Western world. We are still seen to be leading in developing the country on a sustainable basis after the aid dries up." On June 7, 2000, speaking from Pristina, Richard Caborn announced that a consortium headed by British company Npower, and including Alstom and Mitsui Babcock, would lead the refurbishment work at the 660 MW coal-fired Kosovo B power station. These successes were down to the British initiative of a public/private sector partnership that had integrated some of the best people in British Industry with the joined-up government departments of the MoD, DFID, DTI and FCO (through BTI) and DETR.

32

Serbia and Montenegro (2000)

The breakup of the former Yugoslavia into five independent nations had led to the Balkan conflicts of 1991–1995. These were particularly debilitating for the region because the different nations had served as natural resource providers, processors and markets for each other. Among these nations, Serbia was the most highly industrialized—with its industrial sector generating 50 percent of its annual GDP. Serbia's and Montenegro's power generation facilities had been targeted in the NATO bombings of 1999 which destroyed industrial and civilian facilities to the value of some US$4 billion.

In September 2000, I was honored to be appointed a major in the Engineer and Logistic Staff Corps of the British Army. In early October 2000 I was in Brussels speaking at a conference entitled "What Can Business Bring to Balkan Reconstruction?" when some astonishing news came through—the Milosevic regime in Serbia had fallen, overturned by hundreds and thousands of people in a bloodless revolution orchestrated by Vojislav Koštunica[1] and Zoran Đinđić.[2] I was immediately appointed by the British government to chair a task force for Serbia and Montenegro (later to become known as the United Kingdom Government Public Private Task Force for the Federal Republic of Yugoslavia).

At Brussels I'd been talking about my experience of the previous year in Kosovo. I had explained the importance of helping the Kosovars to understand how a market economy works, because many Kosovars reflected fondly on the "pre–Milosevic halcyon days" of the controlled central economy. I had used Kosovo as a model to demonstrate that partnerships involving the private sector are excellent ways of helping major corporations enter the marketplace and contribute to reconstruction. It seemed that I would now be getting the earliest conceivable opportunity to further prove my convictions.

Back in London, in November 2000, I felt deeply privileged to be awarded the Institution of Civil Engineers' International Medal for my work in the reconstruction of Kosovo. At that time Richard Caborn, Minister for Trade, was leading a trade delegation to the Federal Republic of Yugoslavia (FRY) which announced a £1.2 million (US$1.8 million at 2000 exchange rates) package to help in the fields of energy, health, waste disposal and water supplies. DFID was at the same time announcing a £10 million (US$15 million) assistance for Serbia which included humanitarian assistance and technical assistance. The Foreign and Commonwealth Office had already been coordinating a £2.5 million (US$3.8 million) programme of support for democratization in the FRY.

I quickly put together the task force, and we mobilized with the same urgency and speed that had been a feature of the Kosovo task force. We undertook recce assessments to rapidly gain an appreciation of damage and opportunities on the ground in Serbia and Montenegro. On our first task force visit to Belgrade we inspected all the damage inflicted by the U.S. and British bombing. We looked, for example, at the effect of cruise missile attack on building structures. On subsequent visits we investigated the state of the hospitals. We had successes with pragmatic initiatives, such as arranging the delivery of a wide range of medical equipment, from stethoscopes to electron microscopes, for some of Belgrade's major hospitals which had been starved of equipment. We also arranged for help in the repair of the bridge at Novi Sad across the Danube. In another sector we were involved in humanitarian help to provide schooling and housing for some of Serbia's Romani, who were generally living off rubbish tips. Our investigations took in water supplies and environmental assessment and audits (the bombing of chemical factories and oil refineries had caused air, soil and water "hot spots" in Serbia). We assessed the state of the whole energy sector—refineries and petrochemical installations, power generation, power distribution, mining and the associated environmental criteria. We looked at liabilities, existing finance and opportunities for investment. Flagship projects were identified.

The difference this time was that many other initiatives were quickly off the mark. This gave us the opportunity to join our efforts with those of the aid agencies, the European Agency for Reconstruction, DFID, World Bank, European Bank of Reconstruction and Development and, of course, the new government in Yugoslavia (Serbia and Montenegro were not subject to military zoning as Kosovo had been).

The EU was already by far the single largest assistance donor to the Western Balkans. Since 1991, and including 2001, the EU provided more than €6.1 billion (US$5.6 billion at 2000 exchange rates) to the region through its various aid programs. For the year 2001, it made available more than €845 million (US$757 million at 2001 exchange rates) for its existing Phare, Obnova and CARDS (Community Assistance for Reconstruction, Development and Stabilisation) programs alone. In November 2000 the British government lifted its embargo on investments in the economy of the FRY and made the decision to unfreeze Yugoslav financial assets held in British banks. In December 2000 the vast majority of products from South-Eastern Europe were granted duty-free and unlimited access to EU markets—arrangements more generous than those granted to the EU candidate countries of Central and Eastern Europe.

Now the EU committed to providing US$2.5 billion in aid over the forthcoming seven years. The Serbian government negotiated with the IMF, the World Bank, the European Bank for Reconstruction and Development, the London Club and the Paris Club to bring about the normalization of financial flows in the form of credit (these negotiations would lead, for example, to the agreement with the Paris Club, signed on May 22, 2002, which wrote off 51 percent of the Yugoslav debt).

On March 16, 2001, the World Bank approved a US$30 million grant. Germany and Greece pledged to help rebuild the electricity sector. Russia promised natural gas supplies and assistance in repairing damaged thermoelectric facilities. Switzerland supplied electrical spare parts and equipment (circuit breakers, measurement transformers, measurement equipment and protection equipment) as well as a new compressor for the thermo-power plant Nikola Tesla A at Obrenovac. Norway provided mazut fuel oil and diesel fuel supplies.

Whatever the scale of aid and trade initiatives, however, the challenges facing the FRY remained formidable. Prime Minister Zoran Djindjic spelled it out in April 2001 when he said:

> The new leadership of Serbia, which took over in January 2001, is actively engaged in overcoming a number of challenges: to establish as rapidly as possible new democratic institutions governed by the rule of law; to show clearly a commitment to European integration and closer ties with the European Union; to revive an economy in shambles; to offer hope and encouragement to a population brought to the brink of poverty; to revive international economic ties and relations from the lowest level since World War II. A great deal needs to be changed to attract foreign business partners.

Budget deficits under Milosevic had appeared to be negligible but were in fact hidden by his non-payment of international debt. Djindjic had now to contend with a national debt of US$12 billion—a debt of more than 100 percent of GDP. It meant that, while the incoming government set about taking close control of its fiscal affairs, the government deficit as a fraction of GDP would inevitably rise as normal government functions were restored. The economy was expanding, but only by 5 percent (2000) and 5.5 percent (2001) against the very low level of activity of 1999. Milosevic had held the exchange rate at a massively overvalued level, and it was now allowed to depreciate sharply during 2000 and 2001. Inflation was rising from 60 percent (1999) to more than 75 percent (2000) and more than 90 percent (2001)—but it had been as high as 3,000 percent in 1993. Unemployment exceeded 25 percent of the workforce. Much of the employed proportion was inefficiently deployed in overmanned, state-owned enterprises. Public infrastructure was either bombed out or neglected, dilapidated and obsolete.

The downside was very down. But there was an upside. Serbia's market size was potentially equal to that of Hungary and the Czech Republic. Serbian export potential covered many industries. These industries included agriculture and food processing, forestry, publishing and printing, chemicals, rubber, textiles, leather, minerals, iron and steel, metalworking, construction and construction materials, transport and communications. Serbia also offered one of the last investment frontiers of Europe. Investment opportunities ranged across energy and mining, transportation, construction, oil and petrochemicals, manufacturing industry, science and technology, tourism, environmental protection, agriculture and food production. What the FRY had going for it, above anything else, was the collective will of its peoples, its government and the international community. Nobody wanted another war in South-Eastern Europe. There was universal commitment to peace, freedom, stability and prosperity.

The effects of all of us working for reconstruction in the Balkans helped

32. Serbia and Montenegro

Nigel Thompson (right) at a conference discussing Serbia and Montenegro initiatives (author's collection).

shape the policy of the new government in Belgrade. The politicians recognized that business was crucial in securing the future. They took on with enthusiasm the responsibility for creating a business-friendly environment. Their overriding concern in designing institutional change in their economic system was to create a business environment that would be transparent and recognizable to Western businessmen and investors. They set about achieving:

- Trade liberalization.
- A modern investor-led concept of privatization.
- A tax system conducive to investment.
- A Foreign Direct Investment law attractive to foreign investors.
- Accounting and auditing practices comparable to those of other European countries.
- Consolidation of the banking and financial sector.
- A monetary policy that would secure a stable currency and a rapid reduction of the inflation rate.

Because of the economic reforms, inflation in Serbia fell to less than 12 percent in 2003. The Montenegrin government achieved similar success by reducing inflation to 9 percent in 2002—the first time that a single-digit rate of inflation had been achieved in Montenegro since 1976 (with the Belgrade Agreement of March 14, 2002, Serbia and Montenegro had entered into a looser union).

In June 2001 the European Commission and the World Bank, in coordination with the FRY government, had launched a four-year Economic Recovery and Transition Programme (ERTP) of multilateral and bilateral assistance aimed at rebuilding the region's institutions and economy. This outlined a €4.6 billion (US$4.1 billion at 2001 exchange rates) external financial support package needed to reduce poverty and to promote economic stability and growth through political and institutional reforms, the priority being public investments such as health, education and infrastructure, and urgent budgetary and balance of payments support. (According to figures from the World Bank and European Commission, total commitments and planned allocations reached the €4.6 billion target in 2004—international financial institutions provided €2 billion in grants and loans, the EC and EU contributed €1.8 billion and other bilateral donors contributed €800 million.)

UK exports to Serbia and Montenegro were £51.15 million in 2001 (US$73 million at 2001 exchange rates), rising to £61.628 million (US$92.5 million) in 2002. There was now real potential for, and mutual interest in, further expansion of economic cooperation between the FRY and the UK, especially in the fields of energy, agriculture, building construction and telecommunications.

In May 2001 I was delighted to become one of two winners of the newly inaugurated Association of Civil Engineers' Outstanding Contribution Award. The award, presented at the association's annual awards dinner on May 11, was introduced "to recognize outstanding individuals and celebrate the 'unsung heros' within the engineering industry." On presenting the award, the judging panel commented, "Nigel has developed a successful career over many years for a major signature consulting engineering firm. He has undertaken outstanding work overseas—which is of great importance to our profession and future economic success."

On receiving the accolade I said, "Credit for much of the Yugoslav work must go to the task force, which has really demonstrated how well government and the private sector can work together when they try."[3]

On November 17, 2001, Nicky and I were delighted that Sam married Tonia, the second daughter of Levterry and Gaynor Panagiotides of Hayling Island. Sam and Tonia had known each other for a number of years as they were both involved in helping to run huge yachts. Sam had progressed up the ranks from deck hand to master and Tonia from chef to facilities manager. The whole family came to the wedding, including Nicky's sister with her tribe from the Isle of Wight.

The Queen's New Year Honours List in January 2002 included the announcement of a knighthood for my services to international trade and for my work in the reconstruction of the Balkans. In June 2002 I was accompanied to Buckingham Palace by Nicky and my daughters Phoebe and Nim. Prince Charles carried out the investiture in the palace ballroom, demonstrating in our conversation his keen awareness of the situation in the Balkans and of the contracts won by British companies to help with the rebuilding of power, water and other industries in the region. I was sad that my mother was not alive to hear about my knighthood—I know she would have been hugely proud.

I had become a Knight Commander of the Most Distinguished Order of Saint Michael and Saint George. This is a relatively modern order of chivalry founded in 1818 by the future George IV, acting as Prince Regent for this father, George III. Perhaps the most famous member of the order, and certainly the most intellectually gifted, was the Victorian explorer and writer Sir Richard Burton.[4] The order's

32. Serbia and Montenegro

Top and bottom: Nigel Thompson receives from Charles, Prince of Wales, a Knighthood—Knight Commander of the Most Distinguished Order of Saint Michael and Saint George—on the 31st day of December 2001 in the 50th year of the reign of Her Majesty Queen Elizabeth II (British Ceremonial Arts Ltd.).

motto is "Token of a better age," which is about as appropriate as you could get in the context of Southeastern Europe at the start of the twenty-first century.

Shortly after receiving my knighthood, we were thrilled with the news on the August 5, 2002, that Tonia had given birth to a beautiful little girl they named Taia. Our first grandchild—Nicky and I were absolutely delighted. Like Nicky's sister and Tonia's mother, she had wonderful red hair.

33

Stability Pact for South-Eastern Europe (2001)

The Stability Pact for South-Eastern Europe[1] was adopted at a special meeting of foreign ministers, representatives of international organizations, institutions and regional initiatives, in Cologne on June 10, 1999. It established a political commitment to a comprehensive, coordinated and strategic approach to the region. The idea was that as the urgent problems generated by the Kosovo crisis were resolved, the Stability Pact would devote more and more of its time to the longer-term development and stability of the region.

While building on existing structures, the strategy of the Pact was to: secure lasting peace, prosperity and stability for South-Eastern Europe (SEE); foster effective regional cooperation and relations through observing the principles of the Helsinki Final Act 1975 (Conference on Security and Co-operation in Europe); create vibrant market economies based on sound macro-economic policies; and integrate the countries of South-Eastern Europe fully into the European and Atlantic cooperation structures (primarily the EU).

Essentially, then, we were looking at a mechanism for discussion and coordination between the EU and the countries in the region. It was a sort of "forum-cum-clearing-house" for generating ideas and requests for reform-related assistance from the participating nations/institutions. It wasn't a grant-giving agency—decisions on which projects to support were made by International Financial Institutions, the EU, the Organisation for Security and Co-operation in Europe etc. But the idea was that the Pact would be a valuable instrument of change if the countries of the region were to generate the impetus to make it work.

The Pact was inaugurated by the Sarajevo Summit of July 29–30, 1999, which was attended by British Prime Minister Tony Blair and Foreign Secretary Robin Cook. It was set up with 29 participants. These included the countries for the region, EU, Russia and the U.S. Eleven facilitators and five regional initiatives supported its aims and took part in its structures. The facilitators included Japan, Canada and international organizations. Observers included Poland, Ukraine, Switzerland, Norway and Moldova.

A South-Eastern Regional Table was set up as an umbrella body of the Pact.

This was made up of representatives of governments, international organizations and institutions. The Table reviewed progress and provided guidance for advancing objectives. It was organized through three groups, or "Working Tables," which built upon existing expertize, institutions and initiatives in the fields of democratization and human rights (Working Table 1), economic reconstruction, development and cooperation (Working Table 2) and security issues (Working Table 3).

A business advisory council to the Southeast European Cooperative Initiative (SECI) had been launched in 1997, and another such council (BAC) was launched in 1999 to "plug into" the Working Table 2 activities of the Stability Pact. The BAC to the Stability Pact held its inaugural meeting in Berlin on January 19, 2000. Special Coordinator of the Stability Pact, Bodo Hombach, stressed at Berlin that BAC input to the activities of the Pact would be vital because it would assure firsthand knowledge of work on the ground.

At its fourth meeting, held in Istanbul on October 17, 2000, the BAC decided to increase its membership from 20 to 24. This enabled the immediate admittance of representatives from Russia, Spain, Turkey and the United Kingdom (members of the BAC serve as outreaches for their respective business communities). The Foreign and Commonwealth Office invited me, in January 2001, to be the British representative on the BAC. The four new members were, respectively: Valery Otchertzov, Chairman of Intera Holding; Antonia Hernando, Director of Union Fenosa ACEX; Professor Celik Kurdoglu, Chairman of Kurdoglu Consulting; and Nigel Thompson, Deputy Chairman of Ove Arup & Partners.

So I was in familiar territory working across political, cultural and linguistic boundaries to make a difference, while at the same time batting for Britain. Power supplies had been my main concern in Kosovo, in 1999. One of the major concerns of the BAC, at my time of joining, was water supplies.

South-Eastern Europe faced the universal challenge of impressing on its peoples that water is a product that has a cost. In 2000 income from water was too low to cover the operational costs of supplying it. The price of water was about 0.6 DEM/m^3 but, to cover the costs, it needed to rise to around 1 DEM/m^3. The volume of water unaccounted for, because of leakage and unpaid supplies, exceeded 50 percent (compared with the European average of 15–20 percent).

The co-chairman of the BAC was Yves-Thibault de Silguy, who was a member of the executive board of Suez Lyonnaise des Eaux of France. He considered that a 10 percent reduction of leakage could be achieved cost-effectively, within 18 months or two years, by improving the efficiency of the networks. He also flagged up the potential for achieving better customer management in areas such as billing and commercial services.

We were all agreed on the economic, financial, technological and commercial benefits of partnerships between the public and private sectors in businesses like water supply. Such partnerships may range from management contracts to complete privatization. The privatization route is not usually the best option for water. The European Investment Bank could, for example, point out the advantages of successful management contracts set up in Amman and Gaza. The key to successful PPP is to have clear and shared objectives, with the government acting as a facilitator and regulator and the private sector providing the service at the appropriate cost, while respecting the standards and regulation.

For South-Eastern Europe we established priorities of rehabilitating the existing facilities, adapting the institutional frameworks and using private management. We needed these activities to take place concurrently in order to make strong cases for the involvement of international financial institutions.

In the summer of 2001 the Stability Pact created an Infrastructure Steering Group (ISG) to provide international financial institutions with a framework within which they could agree on a policy towards infrastructure development in South-Eastern Europe. This would eliminate duplication, or worse, and make South-Eastern Europe the only region in the world with such a coordinating body among international financial institutions.

A BAC meeting was held at Zurich Airport Conference in October 2002. Chairman Pierre Daures welcomed new member Mrs. Margret Thalwitz, Representative of the World Bank and a member of the ISG. I seized the opportunity to ask Mrs. Thalwitz to provide us with a real picture of ISG projects. This was because the euphoria of the Sarajevo Summit had spawned wholly unrealistic expectations about the scale and speed of infrastructure work. At the Stability Pact First Regional Conference, held in Brussels in March 2000, 34 infrastructure projects had quite unrealistically been included in a list of 244 Quick Start Projects. By October 2002 the list had grown to 46 infrastructure projects, but only 23 had started on site, with a further five imminent.

A joint meeting of the SECI and Stability Pact BACs was held at Skopje, Macedonia, on December 11, 2002. Margret Thalwitz delivered a World Bank Group presentation which included the information on infrastructure projects that I had requested at Zurich just a few weeks previously. The news was good—Margret was looking to add another 12 start-ups in cross-border and regional projects during 2003. I requested that an infrastructure steering group report be presented at every BAC meeting. The reason for the joint meeting was that we were merging the two BACs to form the Business Advisory Council for South-Eastern Europe (BAC for SEE). The BAC for SEE reported to the special coordinator of the Stability Pact for SEE (the SCSP) and targeted eight countries—candidates or candidates-to-be for EU membership—Albania, Bosnia-Herzegovina, Bulgaria, Croatia, Moldova, Macedonia, Romania and Serbia & Montenegro (including Kosovo). I was keen to put accountability processes in place for the new set-up and requested that leaders of BAC activities set down their priorities and specific targets at the beginning of the year. This was agreed.

The merger meant that the BAC for SEE was bigger than its predecessors— we now numbered some 40 countries. Members were active locally, regionally and internationally, with collective experience and networks spanning many sectors including energy, telecommunications, infrastructure and agribusiness. BAC members and collaborating financial institutions and initiatives met four times a year, at their own expense, in main cities of the region, to consider regional and country-specific investment and trade-related issues.

The BAC for SEE aimed to open up South-Eastern Europe to trade and investment and to make the region economically viable and competitive. It recommended new investment projects and presented governments with insight on specific investment considerations. It identified obstacles and disincentives to trade, infrastructure and investment, and worked to eliminate them. Public, private partnerships were promoted where these could deliver functional and competitive solutions. New export and guarantee facilities for the region were encouraged. Support was given for initiatives for good governance and human capital development. Encouragement and support were given to judicial reform and the fight against corruption.

In infrastructure development, the BAC for SEE challenged delays and obstacles created by bureaucracy and the multiplicity of decision-makers across the region. One of the most important outcomes was the realization of cross-border infrastructure projects deemed essential by the business community. We investigated delays in implementing cross-border projects and offered solutions—these activities also helped to encourage and maintain multilateral project financing. We prioritized projects to governments and institutions, and we followed up cases where foreign investors were deemed to receive unfair treatment.

The first meeting of the BAC for SEE was held at Sarajevo on March 19, 2003. It was not a happy start. We began with a minute's silence because Serbian Prime Minister Dr. Zoran Djindjic had been assassinated the previous week. Djindjic with Vojislav Kostunica had, in October 2000, organized the popular demonstrations which had toppled Slobodan Milosevic. As prime minister, he had reformed the Serbian economy to international approval. It was also Djindjic who had ordered the arrest of Milosevic and his extradition to the Hague to face trial for war crimes. It was particularly sad for Serbia that Djindjic had survived years of being targeted for assassination by Milosevic only to be shot after his opponent had been overcome.

Serbia's new prime minister was Zoran Zivkovic who, at 42, was eight years younger than his predecessor. Both men had been founding members of the Centrist Democrat Party so reforming policies would continue. We at the BAC would be able to work well with the new leader because he was a businessman and a pragmatist. In 1999 he had been mayor of Nis, the third-largest city in Serbia, and had controversially accepted heating fuel supplies donated by the EU. When challenged about accepting gifts from countries that had recently been bombing Yugoslavia, he said that Yugoslavs could not afford to wait 15 years to forget the past. I suggested that we at the BAC sent Zoran Zivkovic a telegram, congratulating him on his appointment. At Sarajevo we established "BAC Country Teams" for Albania, Bosnia, Bulgaria, Croatia, Macedonia, Romania and Serbia & Montenegro. The idea was that each country committee would have a chairman from the country in question and a co-chairman from a country representing a major investor. I co-chaired the Serbia & Montenegro country team with Budimir Kostic of the Raiffeisen Bank Jugoslavia, Serbia and Montenegro. I stressed to my colleagues on all the country teams that they should not focus on producing reports. Instead, I said that they should concentrate on issues hampering progress in their defined areas of action. It was a constant of mine that we as an organization should not start to mirror the bureaucracies that we were mandated to fight against.

I had an early opportunity to demonstrate just how I wanted things done with a limited mission to Serbia & Montenegro which was undertaken on June 4–5, 2003. This mission was highly appreciated and established priorities in improvements to the telecommunications network, the provision of facilities for investors and infrastructure works (roads, bridges and intra-regional flights). Budimir and I also identified constraints on development which included a lack of phone lines, river transport and air traffic control. The positive results of the Serbia & Montenegro country team were in marked contrast to the results of teams trying to make a difference in some other SEE countries, notably Albania and Macedonia.

By the end of 2003, I had served for more than three years on firstly the BAC to the Stability Pact and then the BAC for SEE. It was time for me to do my bit to implement the policy of turnover of membership and my resignation was accepted at the BAC for SEE meeting in Bled in May 2004. I was delighted that the bleak days in the Balkans of 1999 had now been replaced by a spirit of optimism and a feeling that the region was entering an era of massive opportunity.

34

The Island of St. Helena (2002)

As an impressionable teenager, I had sailed to a new working life in southern Africa and, from there, back to another new working life in England. During these voyages I first gazed on the island of St. Helena rising up out of the South Atlantic Ocean. Now towards the end of my career, I had a fantastic, once-in-a-lifetime opportunity to help create something of lasting value for the people living in this remote part of the world.

A dream "pub-quiz" question has to be "Name the British Overseas Territories." Such a question would please any landlord by stimulating heated debate, fuelled and lubricated by copious amounts of alcohol. There are, in fact, 14 remaining British Overseas Territories. They are Anguilla, British Antarctic Territory, Bermuda, British Indian Overseas Territory, British Virgin Islands, Cayman Islands, Falkland Islands, Gibraltar, Montserrat, St. Helena and Dependencies (Ascension Island and Tristan da Cunha), Turks and Caicos Islands, Pitcairn Island (plus Henderson, Ducie and Oeno Islands), South Georgia and South Sandwich Islands and the Sovereign Base Areas on Cyprus.

There are some pretty remote places on the list. Of these, St. Helena was chosen as Napoleon's place of exile, following his defeat at Waterloo. Escape from St. Helena, unlike Elba, would be impossible (Napoleon died on St. Helena in 1821). Discovered by the Portuguese in 1502, the island was first settled by England's East India Company and has remained British since 1659. In the eighteenth century it was a vital staging post for ships sailing to the East, with more than 1,000 vessels dropping anchor each year.

By the late twentieth century there were some 4,500 inhabitants—known as Saints—on the island. Their isolation has advantages in terms of quality of life. But there are big disadvantages too. For example, falling seriously ill on St. Helena could necessitate a five-day, 1,500 mile sea voyage to the more advanced hospitals in Cape Town. People who've moved away from the island may put off visits to relatives back home for many years, on account of the time and cost involved. For example, Northamptonshire Newspapers reported on September 25, 2002, that county resident Daphne Brace, 57, was returning to St. Helena for the first time since she left the island at the age of 16, in order to visit her ageing mother Rose. This would also be Daphne's first meeting for 42 years with her brothers Hensel and Colin, and her sister Eileen, who all still lived on St. Helena.

A passenger/cargo ship does sail regularly between Cardiff in Wales and Cape Town, calling at Tenerife, Ascension Island and St. Helena. This ship is the RMS (Royal Mail Ship) *St. Helena*, sailing under the British flag and crewed almost entirely by Saints. The RMS carries a maximum of 128 passengers. Built in 1989 specifically for this route, it is the last in a long series of mail ships formerly operated by the Union Castle Line, and now operated by Andrew Weir Shipping for the St. Helena Line. Because of the distances involved, sea transport cannot be the solution to the Saints' travel needs. The St. Helena Leisure Corporation (SHELCO) was founded in 1987. Its original name was St. Helena Airport Ltd. because, at that time, it was dedicated solely to achieving a practical solution to St. Helena's lack of air access.

Until the early 1990s the British government believed that aviation operations to St. Helena were technically impossible. SHELCO worked to change that perception. They asked me to take on the chairmanship of their company and help develop their case and proposals. I was happy to get involved but asked that SHELCO give Arup first refusal on engineering design and development work. This condition was agreed.

Between 1999 and 2002 SHELCO invested in extensive technical and commercial feasibility studies, involving work by visiting specialists and more than 1,100 man-days on St. Helena. Work undertaken included on-island research, surveys, logistical assessments, infrastructure audits, engineering and architectural design visits, environmental, social and economic scoping, government liaison

2000: Arriving in St. Helena, with Government House/The Castle in foreground (author's collection).

and discussions and public consultations. Airfield study work by Arup covered geotechnical surveys, terrain studies, Moss modeling, runway safeguarding profiles and detailed pavement, and apron and terminal designs. A SHELCO engineering investigation team visit in July 2002 included logistical and other specialists from the British Army Royal Engineers. Having worked with the top general of the Royal Engineers in Kosovo, I approached him about the Royal Engineers doing an exercise on how to land a team of engineers at Prosperous Bay, get all the equipment ashore and scale the 1,000 ft. cliff face to construct an airport runway. The general liked the idea, said they were always looking for suitable exercises for his team and set it in motion. Twelve Royal Engineers duly landed at Prosperous Bay.

The RE studies and our studies enabled SHELCO/Arup to develop and design comprehensive, fully integrated strategic proposals responding directly to the social, economic and technical challenges facing St. Helena. The proposal put forward in 2002 was a package comprising a small international airport to CAA standards, a national St. Helena airline in association with an established, international carrier, and—with a world-ranked hotel group—the creation of an appropriate, environmentally-friendly 5-star resort hotel, spa and golf course (for which rights to a 180 ha/445 acre site had been acquired). The members of

Nigel Thompson signing a Memorandum of Understanding (MOU) with the Governor of St. Helena for providing a hotel, golf course, Wire Bird Sanctuary (The Wire Bird, which does not fly, is unique to St. Helena) and 150 villas (author's collection).

SHELCO consortium and the backers of the consortium wanted to invest initially some of £16 million (US$24 million at 2002 exchange rates) of their own money, additional to the £2 million already spent.

Our package became known as the three-legged-stool proposal. It's a reference to the financial formula, which underpins the airport-airline-resort project. It's good to have an airport, but only if there are guarantees that people will want to fly in and out of it. Any airline operator would have to think twice—and then some—before providing a service for passenger volumes based on the population of a small, remote island. So the leg that locks the stool in position is the resort, which attracts repeat business of people who value the very away-from-it-all quality that St. Helena has to offer. In effect, it's generating and catering for the resort demand that makes an air transport facility economically viable for the benefit of all the Saints. The project as a whole would create some 300 new jobs for the Saints.

As chairman of SHELCO, I was committed to investment in St. Helena, whether for the whole development or, say, just the hotel and resort element. But to split the project would mean losing the values of the synergies that were there for the picking and all the integrated experience built up in developing the package. Also I'm an airports man, and this airport, with the huge social benefits it brings, would be the one I wanted to have as the culmination of my career. During my time with Arup, my colleagues and I designed international airports on every continent together with runways and airports on islands throughout southeast Asia and Australasia. So there was a wealth of experience there to call on, plus current leading-edge skills and world-class capabilities in every aspect of airport planning, design and construction management. Added to that, the contractor that we had on board, Grinaker-LTA, had recently built an airport with a 4,000 m (13,123 ft.) runway on a very difficult site in a part of South Africa in just 14 months. As if that wasn't enough, we have on our team the world-renowned British Army Royal Engineers with everything they have to offer. Once built, the airport would be run by TBI plc, an airport management company that operates many airports around the world.

The type of tourism that we planned on bringing was high-value, low-volume. This corresponded with the formula recommended for the island in the World Tourism Organisation Tourism Master Plan that had been funded by the United Nations Development Programme and carried out in 1996–97. There were no viable alternatives because St. Helena couldn't be expected to sustain, and wouldn't want to sustain, high volumes of anything. And low-volume, low-value would do nothing for the island's economy. So we planned to invest in a 5-star hotel operated by the Oberoi Hotel Group, which runs luxury hotels in Asia, Mauritius and Switzerland. Potential tourists would have the financial means of choosing competing premium holidays throughout the world. So we'd be offering them qualities that they'd have difficulty finding all together in any one alternative destination. This would, in fact, guarantee the future of the island and its tourism industry, because we needed to be able to offer in perpetuity the very attributes that the Saints want to retain in perpetuity—the walking, serenity, beauty, friendliness.

St. Helena Airways would have our aviation director, who's highly experienced in civil aviation and running airlines, as its managing director. The airline would be operating air services from London, Paris and Fort Lauderdale in Florida. We'd be looking at accommodating 50 percent of the visitors at the resort to keep the operation viable. We'd have the capacity to cater for more than 50 percent of the traffic. But, above that 50 percent figure, there'd be alternatives in self-catering accommodation, local guest houses, local hotels, boarding with families, or whatever.

Of course, whenever new building works are mentioned, environmental concerns are expressed. The beauty of our scheme was that we'd be planting a million trees, bushes, and other plants, which would be positive and enhancing for the environment. The Eden Project, a member of SHELCO consortium and world authority on bio-sustainability, would be advising on this project and, indeed, on the project as a whole. We believed that our 18-hole, full-size golf course planned for Broad Bottom would be used as much by Saints as by visitors. Studies indicated that, while tennis and squash facilities on the island were under-used, there was demand for improved and increased golf facilities.

I could go on extolling the virtues of the scheme. But by now you'll be thinking, "If this is such a win-win situation, then why haven't we heard anything on the news, or read anything in the press, about the project being built?" Well, on April 7, 2003, The St. Helena government and the UK Department for International Development issued an "Invitation to Submit Expressions of Interest and to Outline Proposals for the Introduction of Air Access to the Island of St. Helena with Associated Economic Developments." The document outlined, as a requirement for bidders to follow, an interdependent development formula of three principal elements—airport, airline and associated inward investment to fund economic developments. This brief exactly replicated SHELCO/Arup propositions that had been discussed with the St. Helena and UK governments throughout 2002–2003. Partial funding of £26.3 million (US$43 million at 2003 exchange rates) for the airport component of the development only was offered by the UK government, subject to certain conditions being met. The issuing of a public invitation had apparently become necessary in order to comply with government procurement and best value criteria and to assure the British and St. Helena governments that all practical ways of bringing air access to St. Helena had been fully explored. Being placed in a competitive situation was not, in itself, a cause for concern to me. Our work on the island to date had placed us in a very strong position to win any "beauty contest." What did concern me was the drawing out of the preliminary processes, leading to inevitable delays in the realization of the benefits (whomsoever might ultimately be responsible for providing them).

There were very real drawbacks to the "Expressions of Interest" process. For example, the serious airport study work by Arup had been funded by SHELCO, which held the copyright. So respondents to the government's invitation had to be directed for source material to a DFID-funded access study by consultants High-Point Rendel that was based on a six and a half day visit to St. Helena in October 2000 (published June 2001). Of three theoretical runway alignments identified at that time, one had immediately been ruled out by the consultants themselves, on environmental grounds. High-Point Rendel's recommendation, "The Robinson Alignment," failed on safety grounds to meet with the approval of the Civil Aviation Authority. A third option was an early version of the original SHELCO two-runway solution, but without the continuous improvement and development that had taken place between 2000 and 2003.

The closing date for the competition was July 25, 2003. There were four respondents, including SHELCO/Arup consortium. The three other interested parties were construction contractors without airline or international hotel group affiliations. Only one of these three parties sent a single person (himself a freelancer) to visit the island. It was clear to me that only one compliant response had been forthcoming—but there was no feedback from government.

Instead, in September 2003, the St. Helena government announced that it was appointing W.S. Atkins as technical consultant, at a fee of £400,000, to re-evaluate the responses received. Atkins reportedly delivered its recommendations to the British and St. Helena governments in December 2003. Despite rigorous efforts by the St. Helena chamber of commerce, describing itself as "major stakeholders in the consequences of air access and the delivery of new economic development," no information was made available about the competition or the re-evaluation of the responses.

By early 2004 rumors had begun circulating on the island that the British government had temporarily shelved its promised contribution towards the airport, as an economic measure required by the Treasury to divert further aid funding to Iraq and to assist with re-budgeting projects on the UK domestic political agenda. No confirmations, denials or clarifications were made by either the St. Helena or British governments. None of the respondents to the competition received any information about their expressions of interest. In March 2004 SHELCO unilaterally published a 22-page question and answer booklet about its own submission and distributed free copies on the island.

On Monday April 19, 2004, Gareth Thomas MP, the Parliamentary Under-Secretary of State for International Development, issued a written statement to the House of Commons. It stated that the British government, in association with the St. Helena government, had decided to discontinue attempts to develop air access as part of a cross-linked package of private sector investment in which air access would be part-funded by proceeds from other private development. The government acknowledged that air access remained the preferred option of the people of St. Helena for maintaining links with the island after the present ship was withdrawn from service during or soon after 2010. It considered that the costs of providing such air access would likely be "substantial."

This was a setback, but the logic of air access to St. Helena remained inescapable and, more importantly, this was what the Saints wanted. A new air access project team was formed under the leadership of Nigel Kirby (DFID) and Sharon Wainwright (St. Helena government). His Excellency the Governor, Mr. David Hollamby, completed his tour in office in September 2004. He'd been governor for five years but had seen, just six weeks into his term, the need for air access to the island. Now he regretted leaving, feeling that another six months on St. Helena would bring an agreement on air access.

The new governor, His Excellency Mr. Michael Clancy, shared the enthusiasm of his predecessor. We at SHELCO remained fully committed. In February 2005 I arrived on St. Helena with Alasdair Thomson, Chief Executive of SHELCO, and Peter Allport, Chartered Surveyor, and investors and we met up with Joe Terry, SHELCO's representative on the island. We had successful meetings with the air access team, the executive council and the legislative council. Following these talks, His Excellency the Governor and I announced that, subject to a positive airport decision and agreeing acceptable arrangements for an air service, SHELCO and the St. Helena government would work together with the object of achieving:

- A self-sufficient and principally tourist-based economy with, where practical, a Saints-first employment policy.
- Policies to help increase the number of tourists visiting the island, with particular emphasis on generating high-value, low-volume demand.
- The maintenance and enhancement of the environment to ensure that the island could support new development and population growth whilst limiting the impact on its unique ecology.
- The creation and successful operation of a unique flagship resort comprising a 5-star plus hotel, golf course and villas designed to blend sympathetically into the landscape at Broad Bottom and Woodlands.
- The setting up of joint public and private sector arrangements to help promote the island and catalyze the growth of new opportunities, tourist facilities and essential services. The participants, where appropriate, would include St. Helena government, SHELCO and the National Economic Forum.
- If there was a positive decision on the airport, early progress on establishing a timetable. This would include arrangements for an air service and how to enable the flagship resort and related facilities to be completed simultaneously.

On March 14, 2005, Gareth Thomas, International Development Minister, announced the decision that St. Helena would have an international airport. It would be located at Prosperous Bay, on the eastern coast. Here, a 2,250 m (7,382 ft.) runway would cater for long-haul jet aircraft such as the Airbus A320 and Boeing 737–800. The aim was for the airport to be ready to accept flights by 2010, when the island's passenger and supply ship was scheduled to reach the end of its working life.

Immediately following Gareth Thomas's announcement, His Excellency Mr. Michael Clancy said, "I have no doubt that this will lead to economic growth and considerable advantages for the island. The island government and community are most grateful for this expression of confidence by Her Majesty's Government in the future of St. Helena." We signed MOU (Memorandum of Understanding) with the governors, but Atkins persuaded an inexperienced DFID to appoint them to design a new airport and go out to tender in the traditional manner. First, they preselected contractors and, after an initial tender, they selected contractors to negotiate with—Basil Read from South Africa and Impregitio with Arup from Italy.

After lengthy negotiation Basil Read of South Africa was appointed to build the airport with Atkins as project manager. Despite the setback, SHELCO concentrated on obtaining planning to build a new 5-star hotel with Oberoi, and, as the worldwide financial climate had changed, we had to look for finance.

The project went on for many years, and during that time I visited the island on five occasions. Finally in 2018 we sold a controlling interest in SHELCO for £1 million (US$1.33 million) to Paul O'Sullivan's company, the St. Helena Corporation PLC. The existing shareholders of SHELCO have all retained an ongoing interest within a new company which holds 25 percent of SHELCO. The press release reporting on the acquisition of a controlling stake in SHELCO was as follows:

> JOINT MEDIA RELEASE—FOR IMMEDIATE RELEASE—2018-11-06
> SAINT HELENA CORPORATION ACQUIRES GOLF ESTATE DEVELOPMENT COMPANY
> Saint Helena Corporations PLC (SHCPLC) and St Helena Leisure Corporation Limited (SHELCO) are pleased to announce the acquisition of a controlling stake in SHELCO by SHCPLC.
> DEAL PARTICULARS
> Acquisition: By SHCPLC of 75.1% of the issued shares in SHELCO
> Consideration:GBP 1,000,000 settled in cash, with a total project value of >GBP150m over the next five to seven years.
> RELEVANT INFORMATION

SHELCO's only asset is an option to acquire approximately 400 acres (162 hectares) of land on the tropical island of St. Helena. The 162-hectare site has been granted planning permission for a world-class 6,378 yards, 71 par golf estate designed by renowned golf course designers Mackenzie & Ebert. Mackenzie & Ebert have designed well-known courses in Europe, the Middle East, Asia and North America including the Bahamas. The architect for the resort and residential development is well-known London firm, Purcell, whose UK partner Jeremy Blake said, "Having had several meetings with Paul O'Sullivan of SHCPLC, it is clear that he shares our vision and wants to make this project happen as soon as possible. He's made it clear that he will keep all the professional consultants on to make sure he does not lose any institutional knowledge."

As outgoing chairman of SHELCO, I said: "We are delighted to have been able to tie up this deal. Paul O'Sullivan has a long career in property development around the world. Being a British national and having previously served the Crown as a young man, he is ideally placed to take on the task of finishing what we have started. I am particularly pleased that having analysed SHELCO's development strategy for delivering Broad Bottom Estate as part of St. Helena's overall plans for creating a self-sustaining, environmentally friendly economy, Paul O'Sullivan is completely in accord with our development proposals and has agreed to exercise the option before the end of 2020. Paul O'Sullivan has a long and distinguished presence in property development, some of it high-risk, and has proven his mettle in that arena. Apart from his development skills, I admire his obvious determination and staying power which will be required to complete this significant challenge. Although we have relinquished control of our company, the existing shareholders of SHELCO have acquired through the process a minority interest as we all share the aspiration to see St. Helena succeed."

I then said, "When SHELCO tied up the original option agreement just after the turn of the century, they didn't realise they had embarked on a two-decade endeavour to help create a sustainable economy for St. Helena to introduce high-value, low-volume tourism to St. Helena island, which would protect its environment whilst bringing real economic benefits to the Saints."

New chairman of SHELCO and current CEO and shareholder of SHCPLC Paul O'Sullivan said, "This deal brings us a step further to realising our goal of being the principal catalyst to create a sustainable economy for St. Helena on their picturesque island. We are already the single largest, non-government investor on the island, and this will place us in a position to develop an eco-resort with a world-class golf course, which will not be matched anywhere. Having paid the GBP1m to acquire SHELCO, we have also provided for the GBP2.3 m required to exercise the option before the end of 2020, with a view to starting work on site and completing the development, within a further five years. I already have keen interest being shown by two well-known 5-star hotel brands, subject to resolving the direct air access from Europe."

O'Sullivan went on to say, "There has been a lot of negativity surrounding the opening of the airport, which was claimed by some to be a complete failure in the first instance. However the St. Helena government and DFID agencies persevered, and what we have today is an airport capable of meeting the tourism needs of the island well into the future. I am confident in our investment and expect to see profits flowing in about 5 years' time. I am also confident in the government and people of St. Helena and hope they will see this as a further demonstration of our belief in them. Whilst everyone was turning their backs on the island, we were buying land and signing construction contracts. Although we are also pursuing other opportunities on the island, we intend to make this eco-resort with its world class 18-hole golf course our priority. We have already built a luxury three-bedroom villa, as a show-house, to demonstrate what we are capable of, and this will benchmark our quality standards for future buyers."

During the many years that we struggled to achieve a successful conclusion to the St. Helena project, I was greatly assisted by Arup's Steven Luke who had considerable experience in the design of airports and also Malcolm Barrie for the design of all the infrastructure and building works. Our achievements have been to enable an airport to be built for the Saints (though not to our design) and we have laid the foundation for a planned development to encourage high value low volume tourism.

Today South African carrier Airlink operates a weekly service to St. Helena Airport from O.R. Tambo International Airport in Johannesburg, South Africa, via Walvis Bay Airport, Namibia, using smaller-sized aircraft than I'd envisaged back in 2002. Additionally, monthly charter flights operate between Ascension

Island and St. Helena Airport. There are two main airport buildings: the terminal building (incorporating café, gift shop, duty-free shop and restaurant) and the combined building for airport operations (including air traffic control and rescue services).

Whilst in the middle of my work on St. Helena, Nicky and I were very pleased that Nim, having divorced Jeremy, now had a very popular new boyfriend. The whole family were happy therefore when on October 23, 2004, Nim married Andrew Harris at a service in Minal Church followed by a reception at Grove House. Andy trained as a geologist and was working as a project manager for Schlumberger, the international oil exploration company. They had been living together for a while in Nim's new flat in Putney.

35

Campaign to Protect Rural England (2003)

> "Let us try to believe that someday, in spite of all, our descendants will be aware of beauty and amenity and will so live that loveliness is increased in their land because of them, and shall be no longer despised and trampled underfoot."—Clough Williams-Ellis, 1928.[1]

In 2003 I was interviewed for, and was offered, the chairmanship of the Campaign to Protect Rural England (CPRE).[2] The appointment was announced in June 2003. The phrase "poacher turned gamekeeper" cropped up during the interview procedure and afterwards. I made the point that as a civil and structural engineer I was involved in designing and constructing infrastructure, but only after careful consideration of all environmental issues. I cited the view that I was, for instance, in favor of an additional runway for Heathrow Airport. My view was that many of the things I do as an engineer, in, say, developing city infill sites and regenerating urban wastelands, give me specialist knowledge of non-destructive "other ways" in which society can meet its aspirations. CPRE history goes back to the 1920s and to the Clough Williams-Ellis polemic *England and the Octopus*. Although that book denounces the ugliness encroaching all around, it isn't gloomy or depressing. The author passionately advocates quality and sensitivity in design, insists that standardization can be avoided and believes that we can do better in the future than we have in the past. *England and the Octopus* is as relevant today as it ever has been and the author's robust language, unshackled by today's niceties, makes it a particularly enjoyable read (the book can be bought from CPRE).

Max Hastings[3] was asked to take on the presidency of CPRE in 2002 a year before I was appointed chairman. He wrote an article, "The Country Matters," that was published in the *Observer* (June 23, 2002) in the aftermath of the United Kingdom's foot-and-mouth epizootic. Because of its importance, the article provides a context to my appointment as chairman, to work with Max the president. So, as Max is also always worth reading, I'm reproducing the article in full (by kind permission of its author):

> It was another rotten week for the countryside. There was a new foot and mouth alert at a slaughterhouse in Lincolnshire. The National Audit Office delivered a devastating report on fraud by farmers, vets and valuers during last year's foot and mouth outbreak. In west Berkshire, where I live, our neighbouring farmer is shutting down the dairy behind

35. Campaign to Protect Rural England

Nigel Thompson introducing Her Majesty The Queen, Patron of the Campaign for the Protection of Rural England (CPRE), to members of the countryside charity on the eightieth anniversary of its founding (author's collection).

our house. Like most of his kind up and down the land, he is losing a fortune from milk production.

Across Britain, we see the farming industry in desperate straits—and at this time, it is a long-term trend, not a temporary hiccup—while the public's confidence in rural stewardship has never been lower. House builders are intensifying pressure on the Government to release more agricultural land for building, and in many areas they face no resistance from farmers, eager to grab the cash. Laying concrete on fields is by far the most profitable activity in the British countryside today.

I live in the country though, like so many modern country-dwellers, I earn my living from the city. Having given up editing newspapers after 16 years, it seemed about time to do something for a cause other than my own overdraft. When the Council for the Protection of Rural England asked me to become its president, I was flattered to accept.

The Council has been a conspicuous force for good in British life for 75 years. We can have an argument about where to place blame, but none at all about the fact that the countryside is in crisis. It seems a good moment for any of us able to contribute, even on the margin, to get out there and start pumping.

The CPRE's principal role is to campaign for how our countryside looks. Indeed, it plans to change its name from 'Council' to 'Campaign,' because that best reflects what it does. But it is impossible to fight for green fields, hedgerows, and the rights of local communities to resist the relentless cupidity of developers, unless one fights for the economic foundations of rural England.

It seemed right to abolish the name of the Ministry of Agriculture, a production-based body discredited by too many failures. But the creation of the Department for Farming and Rural Affairs[4] is as much a waste of time as renaming the Post Office Consignia,[5] unless it is matched by new ideas. In this as so much else, Labour has a sorry record of failing to match effective executive action to its well-meaning rhetoric.

For half a century, farming was the most command-led industry in Britain. The legacy of food shortage in World War II was allowed to dominate policy for much too long. Today, we can see that the consequences for the environment, for consumers, and even for the long-term interests of farming were disastrous. I made a television documentary, *Cold Comfort Farm*,[6] in 1985. Even then, it was easy to predict most of the misfortunes that have today befallen the countryside. All the sensible pundits could see them coming. Yet governments of all complexions did nothing. Above all, they failed to warn farmers and the public about the mounting crisis.

Today, it is widely accepted that farming must move from an agricultural subsidy system based on production, towards one that emphasises environmental good practice and social support, where this seems justified. But some politicians seem to share with many city dwellers a belief that England can become a huge country park. In reality, while farming must change, it must also remain a profitable activity. Most of the countryside will remain a factory floor for food and natural fuel source production, even if this is conducted on a much more sensitive basis. It will be a tragedy if we end up with empty fields bereft of livestock, which is a serious threat. Our farming neighbour says that he is getting out of dairying not only because he is making no money, but also because he can no longer face the huge burden of regulation inseparable from keeping cows. The CPRE is pressing for further reform of the dreaded Common Agricultural Policy,[7] and for British Government policies which reflect the realities above. It is no longer possible to isolate rural development. We must see the countryside as an integral part of English life in making policy not only for the farming industry, encouraging diversification and environment-friendly activity, but also for transport and planning.

The council is resisting proposals from government further to centralise the planning process. 'Planning is the unsung hero of environmental protection, economic prosperity and the quality of life,' says its commentary on the Government Green Paper.[8] Government's intervention to curtail planning processes was prompted by the length and cost of Heathrow Fifth Terminal Inquiry. Yet fewer than a dozen inquiries since 1984 have lasted more than three months. For the most part, the planning process works pretty well. It must remain the first line of defence against the house builders' lobby.

House builders do a vital job. But it is impossible to mask the fact that their objective is to maximise profit by gaining planning consents for new construction. They have no obligation to consider the interests of the countryside, or even of those who live in it, beyond their immediate customers. Government continues to confuse market demand for housing with housing need. New thinking—and more public money—is required to create 'affordable housing' in rural areas. It is essential to sustain emphasis on brownfield[9] development, while resisting Greenfield[10] building.

We are one of the most over-housed societies in the world, because of our distaste for living in extended families, and commitment to a universal right for each of us to occupy separate homes. Maybe this is what personal choice is about, but the price is high. One of the greatest ironies of English life is that we profess to love the countryside, yet we are killing it by inches.

Perhaps the most important job of the CPRE is to make government, and the public, stop and think before doing things which most blight the environment our descendants will inherit. Opposing relentless development is not nimbyism.[11] CPRE's policy papers offer a coherent picture of what we should be doing, as well as what we should not. As CPRE president, it is not my job to make the organisation's policy. Around the country there is an army of CPRE volunteers, who monitor development, planning, transport and social issues, in the best traditions of voluntary service. The council's experts are then responsible for assessing their merits, carrying out research both local and national, then making a case to government and the media.

If I can be of real use, it is in giving a voice to some of these issues on behalf of the people doing the real work. Non-governmental organisations often find it hard to compete for space in print or on the air against the armoured divisions of Whitehall. Since words are my business, I hope I can provide some ammunition in the struggle.

The English countryside is one of the most fiercely contested environments in the world. It is astonishing that so much of its beauty survives to this day, despite all the pressures. Our business is to see that it prospers into future generations, by showing government and the public some of the ways to avoid inflicting man-made disaster.

Max makes the point that CPRE is not about opposing change. I'm not a fan of mission statements, because they tend to force commonness rather than stimulate differentiation. But we have used our uniqueness to articulate a "reason for being," and it reads as follows, "CPRE exists to protect the countryside as a national asset for everyone, whether they live in town or country. That means recognising the inevitability, and often desirability, of social and economic change. We promote beneficial change in the countryside, the sustainable use of land and good design, whilst taking care to protect the countryside we all value."

In my introduction to the CPRE Annual Review 2005, I translated those general aims into some specific current activities:

As well as campaigning on issues directly relating to the countryside, we are concerned to support the efforts of those regenerating our towns and cities. If our cities don't work, they will continue to sprawl into the countryside around them, eating up our remaining green spaces and increasing people's reliance on cars for commuting, shopping, schooling and entertainment.

So CPRE welcomes many of the principles that lie behind the Government's Sustainable Communities Plan.[12] In particular, we support the aim of regenerating the Thames Gateway. We want to see the area turned into a new type of National Park, regenerating the landscape with improved open spaces and well-designed new housing, particularly affordable housing, built at densities that can support local jobs and services and minimise reliance on the car. But we are deeply concerned about plans for the other growth areas, particularly the London to Peterborough corridor. Building sprawling, low density estates away from the jobs and services on which people depend will both undermine the urban renaissance and destroy our precious open spaces.

Much of the countryside under threat borders on urban settlements. Too often this 'countryside next door' is overlooked because it does not contain spectacular landscapes. It is sometimes rundown and apparently ripe for development, but such open spaces are often deeply valued by those who live nearby and play a vital role in focussing investment within urban areas. This was recognised 50 years ago, when the national Green Belt[13] was established. The time is ripe to revive the spirit that led to the establishment of the Green Belt, an aspect of British planning that is envied around the world....

The important thing about CPRE, as Max wrote, is its members. Working in district, county and regional groups, we use our expertise, enthusiasm, local understanding, strong networks of allies and "ears to the ground" to deliver campaigning locally and regionally. In the most general of terms, our spheres of influence include land use, planning, housing growth, conservation of our heritage of buildings and landscapes, energy and natural resources, transport, food and farming, National Parks and Areas of Outstanding Natural Beauty, and "dark skies."

One of the first CPRE initiatives that I got involved in, on taking up the chairmanship in 2003, was the re-launch of the long-running "Night Blight!" campaign. We have to prevent the spread of light pollution from blotting out the sight of dark, star-filled skies in the countryside. In March 2004 I gave the annual Charles Marques Memorial Lecture to the Institution of Lighting Engineers (ILE), held at the Royal Institution in London. My theme was "Save our Stars." I warned the leaders and experts of the lighting industry that, within a generation or two, light pollution will rob the great majority of Britons of the chance to see star-filled skies in their own country. My messages were more for the reporting media than the ILE, which has published guidance notes on light pollution and supports restrictions on obtrusive light. The thrust of my argument was:

> Young or old, highly educated or totally uninformed, we can all be amazed by the spectacular sight of some 3,000 stars on a really clear night, spread across that dome of inky blue sky.... Nothing else, I would suggest, achieves quite such a combination of beauty and mystery. And nothing else has inspired so much science and religion.... Shouldn't we conserve such things in our own back yard while we still can? I strongly believe that dark skies should be part of the ordinary lowland countryside near where we live, as well as the more remote upland areas.... Countryside which has bright orange or pink skies hanging over it is countryside that has been damaged.

I acknowledged the steps that the lighting industry had taken to address the problem but stressed the need for concerted action right now. I said that we needed to involve the sector's consumers and customers, local and central government as well as environmental organizations such as CPRE. My call to action was reinforced by ILE President Dave McNair.

The "Night Blight!" campaign gained momentum. The government confirmed it would produce an annex devoted to lighting issues when it published its new planning policy statement on pollution (CPRE has argued that the planning system has a key role to play in preventing the spread of light pollution). Also the government's *Clean Neighbourhoods and Environment Bill* was set to make light pollution a statutory nuisance (which had been another campaigning point of "Night Blight!").

At a personal level, associated with dark skies, I host at Grove House a colony of pipistrelle bats. I also put up a barn owl box in one of my barns since when (every year for more than ten years now) our barn owls breed young owls for our valley. Few sights match the drama of a barn owl flying towards you out of a dark sky. Barn owls mostly hunt at dusk and dawn, and their favorite foods are small mammals like field voles, common shrews and wood mice that they forage from hedgerows. Their owlets will be in the nest for about eight weeks before making their first flights, though they'll return for a while to rest during the daytime. CPRE is committed to maintaining hedgerows and all the habitats which make up the interconnected matrix of habitats necessary to sustain biodiversity.

During my four years as chairman, I had the pleasure of visiting a huge number of the CPRE branches throughout the United Kingdom, generally speaking at

35. Campaign to Protect Rural England

their AGM about CPRE policies on affordable housing, green belts and working more closely with the farmers. In my last year I was delighted that, with Max's support, I was able to recruit Bill Bryson[14] to take over from Max as president from May 2007. I was very excited about this because Bill had shown himself as a great lover of the English countryside, the villages and the architecture. In fact, he is on record as saying that he thinks the whole of England should be designated a National Park! I was also pleased that he was determined that more should be done to prevent the appalling amount of litter which was being thrown thoughtlessly from cars and vans without any regard for others or for the quality of our countryside. I knew his popularity with the informed public would be a huge help for CPRE.

As chairman I had many meetings with ministers of the Labour government and also with Prince Charles, who proudly showed a few of us around Highgrove and his farm, where everything was managed to the highest environmental standards. On another occasion I had an interesting morning with Prince Charles as he and his architect showed us around Poundbury,[15] his demonstration of an ideal village next to Dorchester in Dorset. I very much agree with many of the principles behind Poundbury, particularly getting the cars off the streets and mixing affordable housing alongside privately owned, higher-quality housing, without there being any apparent difference.

Having met Sir Jocelyn Stevens, National Chairman of English Heritage, on a number of occasions at CPRE events, I invited him to join me and some of my Arup colleagues for lunch in the Arup dining room. During the lunch Sir Jocelyn said one of his biggest current problems was the traffic on the 303 and its proximity to Stonehenge. Also because the road effectively separated the henge from its main approach and setting. I suggested they should look into a cut and cover tunnel as a cheaper alternative to the expensive bored tunnel. I had previously discussed this idea with an eminent archaeologist who was quite happy with the idea as this would involve a major archeological dig. But the majority of archeologists preferred the bored tunnel approach as they wanted the site to remain mostly undisturbed, though some of my Wiltshire CPRE colleagues wanted the bored tunnel to be significantly extended. When I was still at Arup, I was pleased to see that John Prescot approved the building of a bored tunnel but not the very long tunnel. However, the project was then put on ice.

In 2020 I am delighted to read that the current chairman of English Heritage with HMG support is proposing to proceed with the previously agreed bored tunnel.

Finally, to celebrate our 75 years, we invited our patron, Her Majesty the Queen, to come to a CPRE event at St. James's Palace. I'd had the pleasure of meeting Her Majesty on a number of prior occasions, but this time I had the honor to present to Her Majesty CPRE members from throughout the country plus a few guests, including Monty Don,[16] and to lead her around an exhibition we had prepared. Her Majesty was brilliant and clearly extremely interested in everything that we were doing. She also made it clear that she was not keen to see wind turbines marching across the hills as high voltage electricity pylons do in some areas!

36

The Marlborough Brandt Group and BUILD (2004)

Following the Brandt Commission Report[1] entitled "North, South—a Programme for Survival" in 1981, Dr. Nick Maurice and a few others living in Marlborough, Wiltshire, got together and said that Marlborough, which is a comfortable middle-class market town, should follow the example of the Brandt Report and link with a Muslim African village. After some study they decided on Gambia, being not too far away, in the same time zone and in the British Commonwealth. Gambia first became a Commonwealth member on its independence from Britain in 1965. So Nick wrote to the Foreign Office, DFID and the UK High

Banjul High Street, Gambia (author's photograph).

36. The Marlborough Brandt Group and BUILD

Commission in Banjul. The High Commission responded by recommending Gunjur, a Muslim fishing village, and introduced Nick to the village leader. The Marlborough Brandt Group was then formed. The link is founded on mutual learning through reciprocal visits, a development programme in Gunjur and an education programme in the Marlborough area. The group set up a summer programme whereby a team of enthusiasts from Marlborough, including schoolteachers and pupils, would visit Gunjur for a week, staying with host families and, during the day, helping build additional school classrooms or other similar activities.

The Brandt Group embarked on fundraising to support the activities of the group's building programmes and to enable groups from the Gambia to visit Marlborough. Through Nick's enthusiasm and commitment, the Brandt Group has been flourishing for nearly 40 years. In particular, it gives children and schoolteachers from the Marlborough area life-enriching experiences by, for instance, living with a family for a week in a mud hut with no electric light or plumbing or water from a tap (the water has to be collected from a distant well, usually drawn and brought back to the hut by the wife and carried on her head in a can!).

To celebrate 25 years of success with the Marlborough Brandt Group, Nick Maurice felt it was time to widen the linkup principle and broader its base—hence the birth of BUILD, Building Understanding through International Links for Development.

Members of BUILD: Including (back row, fourth from left) Nigel Thompson and (back row, tenth from left) Dr. Nick Maurice (author's collection).

BUILD believes in a world where "It is not for our own benefit, but for the benefit of our children and children's children that we ourselves should put this world right." (Seretse Khama, 1967.[2])

BUILD believes in a world where the development of living relationships for mutual benefit between the broadest range of individuals and communities in the North and the South is making a widespread, tangible and unique contribution to global peace, security, justice and prosperity. BUILD is UK-based. Its member organizations are AFFORD, All-party Parliamentary Group "Connecting Communities," BBC World Class, British Council, Business in the Community, Cambridge Education Foundation, Council for Ethnic Minority Voluntary Organisations (CEMVO), Commission for Rural Communities, Commonwealth Foundation, Commonwealth Local Government Forum, Commonwealth Youth Exchange Council, ContinYou, Department for International Development, Development Education Association, Eden Project, EKTA-Resource Centre for Women (India), Enable, Fairtrade Foundation, Foreign & Commonwealth Office, Friends of Gaa (Kenya), Friendship North/South (Norway), Gemin-i.org, Humanities Education Centre, Inter Diocesan West Africa Links, International Community of Women with HIV/AIDS (ICW), Just Change (India), Link Community Development, Links Japan, Local Government International Bureau, Marlborough Brandt Group, Mercy Corps Scotland, Muslim Council for Britain, Network of Ugandan Researchers & Research Users (NURRU), Ove Arup & Partners, Oxfam, Partnership for World Mission, Plan UK, Princes Trust, Scouts, Skillshare International, Tropical Health and Education Trust, UKOWLA, UNDP, VSO, Womankind Worldwide.

In February 2005 I made my first visit as chairman of the charity BUILD to Gambia on a Marlborough Brandt Group visit. I was accompanied by Nick Maurice, who is the director of UKOWLA and founder member of BUILD. Traveling with Nick and me were Steve Moran, his 12-year-old son James and others from BUILD. The Moran family had collected school uniforms, following a school closure in Marlborough, to donate to the children of Marlborough's link community Gunjur.

Our party was met at Yundum International Airport by the leader of the Brandt Group Gunjur. Following this we headed out of Banjul towards Gunjur, which lies to the south in Gambia, close to the border with Senegal. It is called Gunjur Village because, although the population is of township size (12,000), it is lacking in roads, infrastructure, running water and electricity. Extended families live together in compounds, with lighting delivered by generators and television powered by car batteries (cars are outnumbered by car batteries in Gunjur). Despite its deprivations, this is a Muslim community in which each and every member considers it an honor to host a visitor.

In Gunjur we were welcomed by the Alkali (mayor). We all stayed with different families who were generous with their hospitality. However it was quite an experience living for a week with no electric light, and, particularly, with the aid of a bucket of water, washing and rinsing off the soapy water in the dark without dropping the soap into the bucket was a nightmare! I was placed with one of the leading families, which had its own house, whereas Steve and James stayed in a family compound. They visited the local schools and discovered some harsh realities. For example, although the schools possess computers that have been donated to them, they can only afford enough petrol to fuel the generators to power the computers for a couple of hours a day. James found that a plus point was his ability to communicate because, although the local children speak Mandinka, they write in English.

Meanwhile, Nick and I had a different agenda to pursue. We held many meetings with different people including the vice president of Gambia, the bishop and a senior imam (to discuss Muslim–Christian relations, with a view to learning from the Gambian example), the British High Commissioner, the director of the UK Department for International Development and the country's Director of Voluntary Services Overseas.

I also did a television broadcast. This was the second item on Gambia's National News. In it, I drew attention to Gunjur's extraordinary relationship that exists between the people of Marlborough and Gunjur. Before leaving the country, the Alkali of Gunjur asked me to convey his best wishes to the Mayor of Marlborough and his thanks to the people of that town for their assistance over the past 23 years.

Anecdotal testimony to that relationship includes the House of Commons Hansard Debates for March 26, 1993. Among the subjects under discussion in the House that day was the Local Government (Overseas Assistance) Bill, Clause 1, Power to provide advice and assistance. The late and much-missed parliamentarian Tony Banks, MP for Newham North-West, said:

> ...it will be clear that parish councils are anxious in many cases to offer overseas assistance. By definition, parish councils are very small. Even the largest employs only a few members of staff. What they lack in staff expertise, however, is more than compensated for in enthusiasm and a direct involvement in the community. In line with the general desire to devolve powers and to involve people down to the lowest level—subsidiarity being the term with which we are most familiar in this place—it is a little puzzling that the sponsor of the Bill and the Minister have not so far shown any great interest in extending the scope of the Bill in that direction. The active involvement of local communities, probably drawing directly on the enthusiasm and expertise of parish councillors and the community, is to be welcomed.

Jim Lester, MP for Broxtowe, added:

> The illustrations given by the honourable Member for Newham, North-West are borne out by my research. Quite by accident I came across such a scheme. Marlborough Town Council has linked up with Gunjur in the Gambia. I would add, by the way, that Marlborough is a small and attractive town. The town council plays only a part in that role. However, it takes over to Gunjur a team of people, all volunteers, who add classrooms to schools and do all sorts of worthwhile work in the Gambia. Marlborough has assisted in the most positive way that I can describe. A young man from that community wrote to me. I did not know him from Adam. He said that he would like to come to this country to study hotel and catering skills and asked whether I could help him. I wrote to a college in London that specialises in teaching those skills. Much to my amazement and gratitude, the principal of that college wrote back to me and said that, as I had asked, he was

prepared to give this young man a scholarship. I was very grateful to him. The problem then arose of what to do next. I had neither the resources nor the facilities to house that young man in London while he pursued his scholarship. Very generously, Marlborough's Brandt Group took on the cost of maintaining him in this country. It also assisted him by providing him with work experience in a local hotel in Marlborough. He has returned to his country with a sheaf of qualifications and recommendations. He is a very enterprising young man. I hope that those qualifications and recommendations will lead to his being quickly ensconced as the manager of one of the new hotels that are being built in the Gambia and providing services in that hotel of a quality that will encourage many people and, I hope, many of my Hon. Friends, to go there for a holiday. In that way we shall assist in developing a very small country, but one which is important to us.

I've concentrated so far on the vibrant relationship between Marlborough and Gunjur because it's a good example of what BUILD is all about—people working together to make a difference for the better in the world. This relationship is also a good example of the two-way benefits that are a feature of every international link. While Nick Maurice and I were visiting the bishop and imam in Gambia, a visit was being arranged for BUILD members to the Muslim and Sikh communities of Birmingham. This was led by Dr. Chris Hewer, diocesan adviser on faith issues to the diocese of Birmingham. The visit included meetings at mosques and at the area's largest Sikh Gudwara, as well as a meeting with staff at Islamic Relief. It provided an opportunity for BUILD members to see at first hand the extraordinary and increasing diversity of the Birmingham population. We learned of the social issues that confront Birmingham and recognized the potential that international partnerships can have an increasing social cohesion in the West Midlands and throughout the United Kingdom. To help communities address the opportunities and challenges of forging links, we commissioned one of our members (the Humanities Education Centre, Tower Hamlets, through its director, Margaret Burr) to undertake the production of a toolkit for community partnerships (Toolkit of Good Practice—Opportunities and Challenges). This has been produced in consultation and collaboration with individuals and organizations in the UK and in the South. It takes the form of topic leaflets on different themes such as reasons for forming a link, the underlying principles, learning for change, evaluation, sustainability and resourcing. Prospective link communities can "mix-and-match" the components of the toolkit to suit their own aspirations and priorities.

BUILD aims to bring global links into the mainstream in the UK to the point that every citizen is touched at some point by international cross-cultural partnership, whether at school, in higher education, through the local authority, town or village, hospital, social, arts or sports clubs, or faith institutions. In pursuit of this aim, we work at a strategic level through the all-party parliamentary group "Connecting Communities" and at an opportunistic level with—for example— the Commission for Africa, the Department for Education and Skills and BBC World Class. We greatly value personal initiative and do much additional work individually through our membership.

Our relationship with the all-party parliamentary group "Connecting Communities" helps frame BUILD policies and stimulates initiatives. At this forum, in

March 2004, Charles Clarke, the then secretary of state for education, accepted our recommendations and said that he wanted all schools to develop a truly international flavor. He said that he would like a higher level of relationship between UK educational establishments, and that the government would provide encouragement through various schemes such as Global Gateway (to help schools find partner schools) and Dreams and Teams (to facilitate sports links). As a result of this meeting, BUILD was asked by Charles Clarke to make inputs to the Department for Education and Science (DfES) International Strategy by helping the department develop a strategy for schools which became entitled "Putting the World into World Class Education."

The next "Connecting Communities" meeting, on May 5, 2004, saw Richard Caborn, Minister for Sport, whom I knew well from our work in Kosovo, considering the role of UK/South partnerships in sport. I asked all those present to say a few words about themselves. The participants included the British Olympic athletes David Hemery, former 400 m hurdles World Record Holder, Olympic gold medalist and member of the Olympic Association, and Bruce Tulloch, who now has 15 years' experience of taking athletes to Kenya and who set up the Kenya Marathon to raise funds. Vincent Bolt, of Action for Children in Conflict, described how he uses sport and drama for therapeutic purposes to help single leg amputees in Sierra Leone. Richard Caborn, a one-time treasurer of the Anti-Apartheid Movement, quoted from Nelson Mandela about the power of sport to connect people. He said that he was new to the idea of North/South linking but was enthusiastic to use the power of sport in international development. At the following "Connecting Communities" meeting Gareth Thomas, under secretary of State for International Development, expanded on the theme by considering the contribution that community partnerships can make to international development.

I co-chaired, with Kevin Barron, MP, the fourth meeting of the group, which was held in the House of Commons on March 14, 2005. Kevin has been involved in the Aston School–Makunduchi, Zanzibar Link and a cluster of schools in South Yorkshire linked to Tanzania. He identifies with the issue of HIV through links to a Village of Hope for abandoned children who are orphans as a result of HIV. Kevin is very prepared to "rattle doors" within government. We used the meeting to seek government commitment to build on three areas of activity within the context of the faith communities: the role of faith partnerships in education, with particular reference to the Anglican diocesan links and their role in bringing learning from partners' dioceses into Church of England schools; the role of faith organizations in social cohesion; and the role of the faith communities in international development.

At the next meeting, on February 28, 2006, Neil Kinnock spoke about the work of the British Council, which operates in 109 countries and is active in another 21. Neil explained the British Council understanding of what makes relationships successful: earning trust—active commitment to mutual understanding; listening to different points of view; and accepting that no one country/individual has the monopoly on wisdom.

I've dwelt on these meetings because of their diverse and significant outcome to BUILD. They provide the opportunity for people from the South to make representation to ministers about the impact of their links on their communities and schools. They facilitate lobbying on particular issues, such as the conflict situation in North Uganda. They enable BUILD to cooperate with the different government departments. They raise the profile of community partnerships at a political level. And they provide individual organizations that are members of BUILD with access to senior politicians.

Because BUILD is all about people, I'd like to close this chapter with a quotation from Sandy Henderson, Senior Programme Coordinator with BUILD member DFID Global School Partnerships: "Our increasingly interdependent world demands greater interaction and understanding between people and greater respect for our differences. More than ever, we must build communities where people of every culture are seen as equal, learn from each other and share a common sense of rights and responsibilities as global citizens."

One of the highlights of my time as chairman of BUILD was to meet and work with Desmond Tutu, a man I had wanted to meet since leaving Rhodesia in 1960.

37

Action for the River Kennet (ARK)

In the early 1990s Jack Ainslie, a local farmer, and Roger de Vere, a famous gynecologist, both keen trout fishermen, had properties in Mildenhall with the River Kennet flowing alongside their land. They put together an action group, including me, who cared about the state of the River Kennet, one of England's famous chalk streams which is fed from a deep aquifer where water has collected from the Marlborough Downs over centuries.

The state of the river was being seriously damaged by heavy abstraction of the water by Thames Water, the UK's largest water and wastewater company. Thames Water was abstracting water from a borehole at Axford and also from the River Og to supply homes and businesses in Swindon and the Kennet Valley. The water exported to Swindon does not return to the Kennet as treated effluent but is lost to the river forever. In 1976, following excessive abstraction, stretches of the river dried up. ARK campaigned strongly for the abstraction to cease, leading to a court case; *ARK versus Thames Water*. Although, inevitably, we lost the court case as they could hire very expensive lawyers, Thames Water realized that they would have to take greater care of England's precious chalk streams. Through constant lobbying over the last 20 years, and by proving that Thames's mismanagement of some sewage treatment works was causing pollution of the river, we persuaded both the Environment Agency and Thames Water to take ARK seriously. I was greatly helped by Richard Aylard, a former private secretary of HRH Prince Charles, who was now a Thames Water Director responsible for public communications. Richard, though working for Thames Water, still assists Prince Charles from time to time on environmental matters.

In around 2005 we realized that ARK had to become more professional. So we hired Charlotte Hitchmough first as an administrator/secretary and then quite quickly as Director of ARK. From that day forward Charlotte has made ARK a successful campaigning environmental charity, principally for protecting the River Kennet and persuading Thames Water to work with us on the improvement of the Kennet. This culminated on July 3, 2017, in a celebration on the completion of a £30 million project to reduce abstraction from the River Kennet and the Og. Thames Water commissioned a new Axford pipeline running 17 km (10.5 miles) from the company's Blunsden reservoir north of Swindon to Whitefield reservoir to the south of Swindon. This latest piece of

ARK volunteers, including ARK director Charlotte Hitchmough (top right), checking the caddisfly to examine the state of the river (courtesy ARK).

River Kennet: tails of the riverbank, with voles returning to the river and its environs (courtesy ARK).

infrastructure will mean less water being abstracted from both the Kennet and the Og.

I had been actively involved in ARK for more than 20 years as a committee member. I was chairman for a couple of years, following Roger de Vere's death, and then finally president. In 2018, my eightieth year, I decided it was time for me to stand down.

38

Governor of St. Paul's School (2007–2011) and President of Old Boys (2015–2017)

In 2005 I was approached at Arup by an administrator from St. Paul's School to enquire whether I would be prepared to join a small working group to review the state of the buildings at the school. Sir David Rowland was asked by the Mercer's Company, the trustees for the school, to form a group to review the master plan and the state and quality of the fabric of the school, which had been totally rebuilt in the 1960s on a new site on the bank of the Thames next to Hammersmith Bridge.

The architect who was appointed by the Mercers in the '60s to prepare the design for the new site chose to use a prefabricated light gauge steel frame system called CLASP (Consortium of Local Authorities Special Programme). This system was developed in 1957 for the rebuilding of schools in post-war Britain. The life span for the system was assumed as 30 years so, not surprisingly, by 2005 the buildings were past the end of their designed life. With poor adaptability and durability, and with very little insulation in the winter, the buildings let all the heat out and, in the summer, the classrooms at times became unbearably hot. Furthermore, the installed insulation was mostly asbestos.

While working at Arup around 1965 I saw that Frank Coffin was working on the new school buildings for St. Paul's. I went to see Frank to look at what Arup were doing for the Mercers. I then found that we were appointed to assist in the application of the CLASP system for an entire new school on the old reservoir site adjacent to the Thames near Hammersmith Bridge. I was horrified because I felt at the time that they should provide something more substantial, better quality and to have a longer life than the CLASP system. I was told at Arup that it was "none of my business and it was what The Mercers wanted!!" So it was ironic that I was asked some 40 years later what I thought. I made my views clear. I thought that the original master plan was very poor, as it took no advantage of being so close to the river, and that the decision to use the CLASP system was a mistake. By 2005 nearly all the schools throughout England which had been built using the system had been replaced. The only reason why the buildings at St. Pauls were

38. Governor of St. Paul's School and President of Old Boys

still being used was because the school had continually spent money on maintenance, new windows etc. Anyway, I joined Sir David Rowland's team which also included Sir Alexander Graham (Mike), who was also a mercer and chairman of the Board of Governors. Both David Rowlands and Mike Graham were at St. Paul's when I was there, David being captain of school during my first year.

When we attended the school, it was in the fine, old Alfred Waterhouse building in Hammersmith. The site consisting of 16 acres (6.5 ha) was originally bought by the Mercers Company as Trustees of the Colet estate in April 1878 for £41,000. The architect Alfred J. Waterhouse was one of the outstanding figures in architecture of the day and was later responsible for the design of the National History Museum in South Kensington. The new building was opened in July 1884. By 1964 these buildings were reluctantly recognized as inadequate, so the Trustees seized the opportunity to acquire a superb riverside site of 43 acres (17.4 ha) only a mile away in Barnes.

As far as our committee was concerned, the trustees had swapped a wonderful building with very poor grounds for a marvelous site with very poor buildings. Anyway, the Mercers accepted our report that the school had to be totally rebuilt, and they agreed that we should look for an architect to prepare a new master plan. Accordingly, we ran a competition which was well advertised. A short list was drawn up, and the selected few were invited to prepare designs from which the architect would be chosen. I enjoyed being on the client side and helped select Nicholas Hare as the architect.

Once the architect, Nicholas Hare, was appointed as architect and master planner, we had to select engineers—civil, structural, electrical and mechanical—to work with the architect. Obviously, I had to stand down from the selection process as Arup were, inevitably, on the short list of suitable engineers. Arup were duly appointed as the multidisciplinary engineers for the project.

In 2007 Sir David Rowland stood down as chairman of the redevelopment committee. I was appointed a governor of the school and chairman of the project steering group for redeveloping St. Paul's School.

During the next four years I had the privilege of steering the design and the construction for the first phase of the new development, which comprised new infrastructure and a new science block of 18 state-of-the-art laboratories (six for physics, six for chemistry and six for biology), new porter's lodge, administration facilities and a new cloistered court. It was also a very interesting time to be a governor.

One of the first issues I had to deal with was that the school had built a new building on land that was owned by Thames Water without their permission. Having discovered this, I approached my friend Richard Aylard at Thames Water. They were happy to help us as they needed to ensure that what we were about to do in the rebuilding of the school on the St. Paul's land would not inhibit Thames's plans for building a reservoir sometime in the future under the main cricket and rugby playing fields. With the help of the clerk to the Mercers and Richard Aylard, we were able to sort out a number of legal issues.

The high master who started the redevelopment was Dr. Martin Stephen. When he stood down as high master, we had to find a replacement, which involved the board of directors in many interviews. We finally selected the brilliant Professor Mark Bailey. who became an outstanding high master. It no doubt helped that he was very popular with the boys, the common room and the parents. Besides being a professor with an honors degree from Cambridge, he had a blue for both rugby and cricket and also played rugby for England. After many difficulties we managed to complete the building of Phase 1. The architecture was greatly admired by the school, parents and old boys. I was particularly pleased to have given the school 18 fantastic laboratories and a modern court reminiscent of Oxbridge colleges with a view to the river. A new chairman of governors was appointed, and he decided to have a change of architect for the next phase. He also decided that, as I was now over 70, I should also stand down.

On standing down as governor I was invited on to the committee of the Old Pauline Club. In 2015 I became president of the club, and it was helpful that I had been a governor as I was able to improve the relationship between the Old Boys, the Mercers, the governors and the school. It was necessary as the relationship had deteriorated with the most recent chairman of governors. However with his replacement, Johnny Robertson, and with the help of Mark Bailey, the newly appointed high master, the relationship with all parties was easily and quickly restored. Being president of the Old Boys Club meant that you were also invited to a number of dinners at both the school and at the Mercers. Probably the most important event of the year for the Old Boys is the Feast Dinner held in Mercers Hall after a service at St. Paul's Cathedral where the president reads the lesson. I really enjoyed this fantastic honor, standing under the dome in front of about 300 people reading the lesson. Where my words echoed around the huge space. It was one of those occasions when I could not help thinking how proud my mother would have been and how she would have loved to have been there.

I completed my time as president in 2017. Another of the events that I enjoyed and am also proud of was in May 2017 when I was asked by Johnny Robertson, chairman of the board of governors, to be the apposer at apposition. The reason for apposition is for the governors to review the last year's performance of the high master by judging declarations from a few boys. This is also the day in the school year when prizes are handed out. It is only for the very bright boys—when I was at the school, I was never invited to apposition! The apposer on this occasion had to review four declarations by boys in front of all the boys, their parents, the governors, high master and all the masters and then hand out the prizes. I am attaching below my review of the four declarations.

St. Paul's School Apposition 2017

The Four Declarations

- *How to Flip a Water Bottle* by Lawrence Tray.
- *Weakness of Will* by Adam Rachman.
- *Is CRISPR Gene Editing Really an Ethical Game-Changer?* by Will Saunter
- *Free Trade* by Daniel Whitham.

Before I start I have to admit that, unlike you, I received drafts of the four declarations, enabling me to seek help should it be necessary!

I would like to begin with the declaration on Genetic Engineering.

Is CRISPR Gene Editing Really an Ethical Game-Changer?
by Will Saunter

A brilliant and crisp introduction to the new gene editing technique that has swept through biological research. Will has explained what makes it so exciting to researchers—namely, how the accessibility and speed of CRISPR makes genetic engineering available as an everyday technique to be used throughout the world.

He touches on the potential CRISPR-based research for cures that are being developed for cancer, HIV, Malaria as well as Alzheimer's, Parkinson's and Huntington's.

Will briefly mentions the extraordinary proposal to tackle HIV infection by cutting out the viral DNA infected into the chromosomes of infected cells. From a scientific point of view, it is undoubtedly exciting and offers the theoretical possibility of an actual case, with complete removal of the viral DNA from the body.

Some of the purported approaches involve gene divides which cannot be recalled once unleashed—killing mosquitoes with a CRISPR-based gene divide is not the same as spraying with DDT which can be halted at any time. If successful, and malaria transmission is halted, how do we know whether we will come to regret eliminating a species from the ecosystem? Beware the rule of unintended consequences.

Finally, Will explores more of the ethical issues raised by the new possibilities, but acknowledges that this is clearly a huge area of complex debate which is only just beginning.

In this outstanding declaration I believe Will has demonstrated that although CRISPR has certainly revolutionized medicine and given us a powerful tool it has not transformed the ethical debate on gene editing—or as he says—NOT YET!

Daniel Whitham's Declaration on Free Trade versus Protectionism

This is particularly topical and relevant to the debates which we have all faced during the election and negotiations over Brexit.

I think Daniel has articulated and very clearly identified some of the difficulties and contradictions which exist between free trade and protectionism—highlighting the human tendency to prioritize short-termism. Interestingly, our government may wish to conclude a fair-trade deal with the EU (beware the Wallonias!).

Whilst our farmers will be strongly against offering fair trade deals to Australia, New Zealand and the USA, as they fear being flooded with cheap lamb, butter and grain, those same farmers want fair trade with the EU!

What's wrong with the WTO rules? Is no deal better than a bad deal? I think Daniel should offer his services to Her Majesty's Government!

How to Flip a Water Bottle *by Lawrence Tray*

Next we come to Lawrence Tray and flipping a water bottle!

Not being a follower of social media, I had not heard about this new craze until I spoke to my daughter and grandson. So I then went online and saw a video which dismissed it as a trick and made me wonder at the school and the high master!

But no, it isn't a trick. As Lawrence has so clearly explained to us. He has carried out some quite remarkable analysis into something that most would dismiss as mere trivia. There's some real physics behind what's going on; both in how the bottle slows down at the start and how it's possible to increase the odds of pulling it off.... Apologies, high master!

Lawrence asks if we were at any point wondering if his work had any relevance. I certainly could not think of any! However his physics teacher told me that he couldn't help but be reminded of a story Richard Feynman—another great physicist—tells about how he came up with a thought that would ultimately win him the Nobel prize, having observed some students spinning plates in the campus cafeteria.

In his eccentric collection of autobiographical stories, Richard Feynman recounts: "I was in the cafeteria and some guy, fooling around, throws a plate in the air. As the plate went up in the air, I saw it wobble, and I noticed the red medallion of Cornell on the plate going around. It was pretty obvious to me that the medallion rotates twice as fast as the wobble rate—two to one. It came out a complicated equation! I went on to work out equations for wobbles. Then I thought about how the electron orbits start to move in relativity. Then there's the Dirac equation in electrodynamics. And then quantum electrodynamics. And before I knew it ... the whole business that I got the Nobel prize for came from that piddling around with the wobbling plate." A replica of the Cornell plate is now part of an exhibit marking the centennial of the Nobel Prize. This is taken from "Wolfram.com."

Like Feynman, Lawrence is someone who can make science more approachable and interesting to the layperson. And I commend him for this. We don't know what interesting things might further come out of his analysis if he continues to think in creative ways and play, as I'm sure he will.

And as to his struggles with the mathematics of the bottle in flight, I am surely confident that he will return to this and get there one day. Well done.

Weakness of Will *by Adam Rachman*

Though clearly presented, I found this declaration the most difficult. Doing some philosophical analysis of an everyday occurrence—eating a slice of cake!

I have always been clear about what I wanted to do. When I was 13 at St. Paul's, I had decided I wanted to be an engineer so, from class 4b, I advanced to 5x then 6x. I do not really understand the weakness of will bit! Of course, I am too fat—so I don't eat a lot of cake and I don't worry about it! Being an engineer and latterly a hobby farmer, I am not the right man to philosophize over eating cake.

However, I like the proposition by Daniel Kahneman—*Thinking Fast and Slow*—System one—instinctual and impulsive—System two being logical, conscious and more labored. Clearly, I will have to spend more time looking at my sheep and thinking....

Adam, what do you think are the reasons why some people are weaker of will than others?

Nigel Thompson

May 4, 2017

It will be no surprise that the high master passed his apposition review.

39

Minal Parish Council (2000–2018)

Having lived in the hamlet of Stitchcombe in the parish of Mildenhall (known locally as Minal) since 1972, I was approached in 2000 by Miss Pat Courtman, the formidable chairman of the Parish Council, to join the council. Now that I was home most evenings I had the time to attend the bi-monthly Parish Council meetings and to help in other issues—particularly in planning. Miss Pat Courtman was the only daughter of the old vicar who was resident vicar in 1972 when we moved to Stitchcombe—he was one of the old-school vicars and still owned "the living" for life, which meant that, while he was alive, he kept all the money from the collection each Sunday. He was a huge character and drove jerkily around in an old Morris Minor.[1] As Pat was his daughter, she had her own special seat in the church and, by 1972, chaired both the PCC (Parochial Church Council) and the PC (Parish Council), which she had chaired since 1967, having joined the PC in June 1952. She died in May 2004, having been a parish councillor for 52 years. Parish Council meetings were remarkably similar to *The Vicar of Dibley*.[2] I was asked by the councillors to take over as chairman, having been her deputy for the last two years. We had a clerk who took hopeless minutes—he was a very dry Scotsman from Glasgow and probably the only socialist in the village, which was a nest of Lib-Dems![3] His dry wit kept us all amused. Pat had worked tirelessly for Minal, for the residents, for the wellbeing of the village and the church. A very good example is Church field—the land immediately to the north of the church alongside the graveyard. Pat held a tenancy with the Church Commissioners for the field, where she kept her ponies, and the commissioners offered her a substantial sum to give up her tenancy so that they could sell the field for development. But Pat refused because she believed it would be in the interests of the village if the Parish Council could purchase the field with her as sitting tenant, so that in the future part of the land could be used for an extension to the graveyard and also for the use of the village, and for the church to be protected in its rural setting. With the help of Tom Ottley I was able to negotiate with the agents for the commissioners to purchase the field for the Parish Council and to arrange for an extension to the graveyard.

Having taken over from Pat, I was then chairman for 15 years until Spring 2018, when I thought it was time for someone younger to take it on. I think we did quite well as a council, winning the best-kept village award on three occasions, but I am particularly proud of raising the funds for the design and the construction of a new super playground for the village. It is recognized as being one of the best playgrounds in Wiltshire. So many families come to visit the playground that we had to purchase land off the neighboring farmer to provide some additional space for parking.

40

The Trevor Estate
(1976–2011)

Trevor Square is an elongated garden square in Knightsbridge, London. It was designed in the 1810s chiefly by architect William Fuller-Pocock. The mid-rise basemented houses fronting its two long sides, with slate mansard roofs, are listed in the British protective and recognizing scheme, and were built in the 1820s. The main stonemason employed was Lancelot Edward Wood, after whom is named neighboring Lancelot Place (originally Petwin Place).

The square is in the west of Knightsbridge, north of Brompton Road. In Anglican terms it is in South Kensington but, historically, the parish near-equated to the detached part of St. Margaret's Westminster, which explains the present borough. Arthur Hill-Trevor, third Viscount Dangannon, had agreed to demolish his Powis House in 1811 to make way for the new development.

In 1816 the Rev. John Morrison established the Trevor Chapel on the corner of Lancelot Place. This closed in 1902, to be replaced by a Harrods showroom. By 1911, the minor, the overarching legal interest (reversion) of the square's garden and any buildings enjoyed on long leases (by others) belonged to J.C. Humphreys, Nicky's great-grandfather (grandfather of Nicky's father on his mother's side). The freehold of all the houses surrounding the square (the east side adjoining Trevor Street, the frontage on to Knightsbridge, Hill House and Lancelot Place) was owned by J.C. Humphreys. He sold Lancelot Place and gave long leases to the houses in Trevor Street, the Knightsbridge frontage and Hill House. The houses surrounding the square he let to Harrods on another long lease.

J.C. Humphreys was a barrister and led an interesting life. He had two families each with two children, a boy and a girl. One family lived with him on the Trevor Estate while the other family, unknown to the first, lived also with him in Islington. To keep his secret he gave the same name to the children of each family. The deceit only came to light after his death!

Nicky's grandmother, whom she never met, was the daughter of family number one. Her brother was the father of David Humphreys, also a barrister who, after his father died, took on the role of managing the estate. J.C. Humphreys left 52.5 percent of his shares to his son (David's father) and the remaining 47.5 to Nicky's grandmother. When David's father died, he left his shares to David, David's wife Julia and their three daughters, Belinda, Julia and Sophie. When

40. The Trevor Estate

Nicky's grandmother died, she left her shares equally to Eustace (Nicky's father), Trevor and Cynthia.

In 1976 I was invited by Nicky's father, who was constantly in dispute with David Humphreys over the management of the estate, to join him at a meeting with the directors of the estate to discuss the future. David wanted to get rid of managing the estate but would not let Jim (Eustace—Nicky's father) take over. One of our successes was when I was able to help Jim persuade David not to give the whole estate away to Save and Prosper, an insurance company which had tempted David to part with the estate for a very modest figure.

In 1985 Belinda Howard, David's daughter, and I were invited to join the Trevor Estate board to represent the younger generation. By this time the 47.5 percent of the Bonnett shares had been distributed to 25 people—Eustace, Trevor and Cynthia, Jim and Bridget's five children and fifteen grandchildren and me in the capacity of director.

David was clear that there were now so many family members that it was becoming increasingly difficult to agree policy and manage the estate—most wanted to break the estate up and receive their inheritance. This was particularly true for the grandchildren because they were all of an age when they needed help in purchasing a house. As most of the properties had been let on long leases, some of the tenants were able by law to buy their freeholds, and the others, as sitting tenants, were able to purchase their freeholds also at a very low rate. We were finally able to sell the freeholds, pay the capital gains tax and corporation tax and then distribute the remaining funds according to the shareholding. There were 6,000 shares in total. The Bonnett side amounted to 2,850 shares (47.5 percent). The total received after liquidation of the company in January 2011 was of the order of £15.9 million (US$25.5 million at 2011 exchange rates) which, if distributed to the shareholders evenly, would amount to about £300,000 (US$480,760 at 2011 exchange rates) each—a huge help to anyone wanting to buy a house.

41

Grove House and Estate

Grove House in the hamlet of Stitchcombe, at Mildenhall in Wiltshire, stands on land which in 757 belonged to Cynewulf, King of Wessex. About 786 Cynewulf gave it to his thegn,[1] Bica, who granted it to Glastonbury Abbey. At the taking of the great Domesday survey in 1086 it was held of the Abbey by Edward of Salisbury. In 1275 it was the property of Margaret Longpee, wife of Henry de Lacey, Earl of Lincoln, who held it in her own right. The site of Grove House, together with other land within the manor of Mildenhall, passed to Margaret's daughter, Alice, Countess of Lincoln and of Salisbury who in 1325 granted it to Hugh le Despenser, Earl of Winchester. After Despencer's execution in 1326 it was confiscated by the Crown. In 1342–3 it was held by Reynold de Mohun and Isabel, his wife. It continued with the de Mohuns until 1352. In 1362 it belonged to Sir Thomas Hungerford. On the attainder in 1540 of Sir Walter Hungerford, it

Grove House, Stitchcombe, circa 1910 (author's collection).

again fell forfeit to the Crown. Henry VIII granted it first to his fifth queen, Catherine Howard, then to his last queen Catherine Parr. It was later restored to the Hungerfords. In 1731 it was sold to Charles Bruce, Baron Bruce, afterwards Earl of Ailesbury. It subsequently passed with that title until George Brudenell-Bruce, Marquis of Ailesbury, sold the Savernake Estate in 1929–30 when Grove House (then called Grove Farm) was sold to Mrs. L.M. Edwards.

These, then, have been the titular owners of Grove House and site through some part of fourteen centuries. What of the house itself, and those who have lived in it? Although English Heritage, when listing it Grade II as a dwelling of "architectural and/or historic interest" described it as "eighteenth century," the property has a much older nucleus, as records confirm. The Queen Anne front was added about 1710. The oldest section was put up about 1550, the third year of the reign of the boy-king Edward VI, for the purpose of increasing the manorial rent-rolls through letting to yeoman farmers. The term "yeoman" was first noted in 1362 by Piers Plowman and probably meant "young men."

Yeomen were the backbone of middle England: conservative farmers and "freemen," fiercely proud of their independence from feudal servitude (at least from Tudor times). A yeoman had his own servants and was addressed as "Master," his wife being called "Mistress." From these terms derive the modern "Mr." and "Mrs." They would have held Grove Farm not for a term of years but for a series of lives—most commonly their own, their son's and their grandson's. This

Grove House, Stitchcombe, 2020 (author's collection).

interest was termed "copyhold" because details of it were, literally, copied into the Manorial Court Rolls. It was quite common for the chief copyholder to issue subleases of his interest from which underleases could sprout with such complexity that it is sometimes almost impossible to say who the physical occupant of a property might have been. In this case we know that in 1569, the eleventh year of the reign of Elizabeth 1, the copyholder was Thomas Cooke, yeoman, whose will was proved January 28, 1577. We know from this that he owned or held a copyhold interest in property in the Borough of Marlborough which he bequeathed to his son, Thomas, and "the heirs male of his bodie." To another son, Richard (in the will called "Ricardo") he left "twentie poundes of lawfull money," half of which was paid to him "within a year of my death," and the remainder a year after that. A third son, William, received a similar sum with similar conditions attached. To his daughter, Anne, he left "sixtie poundes." Various relations and friends each received "one yarlinge bullock."

The next copyholder to have come down to us by name is Hugh Chandler, whose will was proved March 12, 1629. Chandler seems not to have been as rich as Thomas Cooke. To the upkeep of the parish church at Mildenhall he left 6s 8d, but "twenty shillinges" to "Mr. Sheatt for a sermon to be preached at my funerall." To his five sons and two daughters he left seven pounds each, the equivalent in modern money of perhaps £650 (US$870), to be paid to them "when they shall be of the age of Sixteene yeares." To his eldest son and heir, John, he bequeathed "my best cloake and my Bigger Bible." His second son, Hugh, received "my second cloake and two silver spoones." The third boy, Edward, got "my other cloake" and Robert "my littler bible." His daughter, Anne, received "my goulde ringe which I used to weare," and Dorothea received his best set of pillows. His sister, Mary Toft, was granted an annuity of "twentie shillings yearly," while her four brothers and four other sisters each received sums of five shillings.

We know that in the 1670s the landholding of Grove House included 82 acres of pasture in small closes, common pasture for 400 sheep and 11 acres in the open fields. The copyholder at this time is not known. By the time of the first Jacobite rebellion in 1715 the farm was in the hands of Thomas Neate (sometimes mistranscribed as "Neale" in records) whose grandson, also called Thomas, held it at this death in 1792. "Thomas Neate of Grove" left a will of admirable shortness, but one that seems to have played his children off one against the other: "I give unto my son Edward Neate one shilling and to my daughter Lucie one Chest and Trunk and one hundred and fifty pounds and to my son Thomas fourty pounds and to my son Robert one hundred pounds and to my daughter Anne, one shilling, all my legacies to be paid in Twenty Months after my decease...." His eldest son, John, received the balance of his father's estate including the copyhold interest in Grove Farm.

From the end of the Napoleonic Wars in 1815 until 1837 Grove Farm disappears from records. It resurfaces at the drawing-up of the tithe map in 1838 when the occupier was shown as Edward Vaisey and the owner as Lord Ailesbury. The "farmhouse, yard, garden and outbuildings" covered two rods and 25 perches.[2] Vaisey farmed a total of 123 acres of arable and pastureland in the parish including Boulcher's Ground (18 acres) and Mildenhall Field (65 acres). He continued here until his death circa 1846 when the copyhold interest devolved upon his son John, who, at the taking of the 1861 census, was occupying the farmhouse, aged 55. He described himself as "farmer of 550 acres employing 24 men and 8 boys." He shared the farmhouse with his wife, Ann, 52, his unmarried brother, Henry 47, and his nephew James 17. The family was cared for by a single resident domestic, Emily Wheeler, 28. We should spare a thought for Emily Wheeler down the tunnel of the years. As the only general domestic kept, she would have been responsible for all the chores outside the kitchen—and a good many within it—including the carrying of coals and bath water to the top of the house and the lighting of all the fires (a thrifty and dexterous servant was expected to light a fire with six pieces of kindling).

By 1867 Grove and Mildenhall Farms had been merged as one arable farm (they separated at the sale of 1929–30). The Vaisey family gave up the tenancy shortly after the death of Thomas Vaisey, about 1877. A new tenant was found in the person of John K. Sargent who, in 1881, aged 36, was farming 600 acres with the help of seven men and four boys—about a third of the labor force employed by John Vaisey to farm a similar acreage 20 years earlier (in the light of this, it is not surprising that, when an estate survey was carried out for Lord Ailesbury in 1867, John Vaisey's Grove Farm was described as "badly managed").

Before 1888 the tenancy passed to Joseph Bathe Cundell, 32, who in 1891 was living here with his wife, Margaret, and their four young children. In 1895 Cundell disposed of his tenancy to William H. Chambers. The consideration paid is not known but, as part of the transfer, the firm of Lavingtons, auctioneers, was asked to carry out an inventory of the property. This survives in Swindon and Wiltshire History Centre in the form of penciled entries in a red leather notebook. Truth be told, Cundell left very little to catalogue—a six-peg hat rail in the kitchen, a copper in the brewhouse, a safe in the pantry and seven roller blinds in the principal bedrooms. The dilapidations were settled at £273 3s 7d on Christmas day 1895.

William Henry Chambers farmed here until his death circa 1906. The 1911 census shows the farmhouse in the occupation of his widow Lucy, 63, who was farming here in her own right. The house then had 11 principal rooms including the kitchen. The Chambers family continued to farm here until at least the outbreak of the Second World War. As mentioned above, when the Savernake Estate was broken up in 1929–30, Grove House was purchased by Mrs. L.M. Edwards, the Chambers' landlady. When she died in 1946 the farm passed to her son R.H. Edwards. When he in turn died in 1971 the land was acquired by Geoffrey Young, who sold most of it sometime later to Mr. and Mrs. J.S. Burrows. In 1972 Nigel and Nicola Thompson acquired the house, since when it has also been the home of their children, Naomi, Sam and Phoebe.

Sir Nigel and Lady Thompson are thus the latest in a long line of owners and occupiers of this lovely old property and site spanning more than twelve hundred years ... from the days of the eighth century when this land was farmed by Bica, a thegn of Cynewulf, King of Wessex.

In 1972 we bought Grove House and the 13.5 acres immediately around the house. In the Autumn of 1999 the Burrows put the 200 acres of Grove farmland including the new farmhouse and dairy on the market. We were approached by Alastair Ewing to see if we would join the Ramsbury Estates and jointly buy the farm because they only wanted some of the land and the woods north of the dairy. They hoped we would want the remainder being the farmhouse, dairy and about 50 acres. I said "yes" but I did not really want the house or the dairy. Martin Gibson, who wanted to buy the farm, heard that we were in partnership with Ramsbury Estates. So he approached me and said that, if he let me have the buildings and say 5 acres alongside our land, would I ditch Ramsbury Estates and join him. I agreed, so in April 2000 we bought 5 acres off Martin Gibson, whose offer had been accepted by the Burrows. This was the land and buildings behind and alongside our stables.

In June 2014 I finally secured another piece of land, a small wood of about half an acre on our Eastern boundary. This small wood had been excluded by Geoffrey Young in our initial contract because he wanted to keep it for access from the main road for managing his shoots. I had discovered that when he sold Grove Farm to the Burrows, this small wood was not included in the sale. I was able to demonstrate to the Land Registry that I had the sole occupancy of this piece of land for the last 40 years. This meant that we now owned circa 19 acres.

After our initial major refurbishment of the house, where we removed the internal staircase and cantilevered a new staircase in a glass tower immediately opposite the main entrance, we only made three further alterations a few years later. Firstly, we installed a steel, circular staircase outside and to the west of the kitchen, going from ground level up to the second floor so that the top floor could be used as a flat. The second alteration was when we had the vineyard because we needed a cellar in which to lock up and store the wine for legal reasons, so we dug down outside our East wall and made an external entrance into the cellar. The third house alteration was to build a conservatory on to the West side of the house. Sometime after we gave up the vineyard Nicky gave me a full-size snooker table for my sixtieth birthday. It fitted nicely into the large presentation room over what had been the vats and fermentation room, which then became a garage!

During the last 48 years we have made many changes to the land. The first major change was when sadly the Grove, which was all elms, developed Dutch Elm disease, necessitating the removal of the entire wood.

Initially we used the paddocks for ponies. At the same time, we had a period when from Matilda, our large white sow, we had a whole series of piglets which often used to escape. Nicky's next venture was to have orphan lambs for the children to help with the feeding. But in no time the parents had to do it all, particularly as—when the lambs grew—they tended to butt you to receive the milk from a bottle!

We then had the vineyard and also, for a time, we let one field to Sonia Wright to run a flower nursery. When we started the vineyard we converted a cattle shed into two cottages for vineyard workers, we created the cellar and we built a courtyard of buildings in which to make and store the wine in huge vats. As I wrote earlier, we also moved the granary from one side of the house to the other to act as our shop.

After the vineyard we had Nicky's Arab horses. We extended the stables so that we could have seven in livery. All the fields had to be well fenced.

During that period, from 1972 up to Nicky's accident in 1992, we always had a least one dog—mostly Labradors. First we had Oscar, and then we had Tosca, who had two litters of beautiful pups. We then had Henry, who was the Labrador that hung onto Nicky's helmet when she had her accident.

After Henry we had Lenny (Lenny Henry!). With Lenny for a short time we had Bertie, a border terrier who was a ferocious hunter, and, with Lenny, had, on more than one occasion, caught and killed a muntjac. Sadly, Bertie was killed crossing the A4 in pursuit of a hare. He had been out walking with Nicky but disappeared when had put up the hare.

We then had Marco and Pollo, brothers but which, not having a good field-trial background, were not a success. Next came Freddie, a great success, who was joined by Cleo, our first sandy/fox-colored Labrador bitch. All the previous Labradors had been black.

After 1992 we sold the Arab horses and let the fields and stables to a number of girls. With the additional land and buildings we bred geese for a few years before going back to chickens. Although there had always been chickens at Grove, previously they were owned by Hilary, who we allowed to use first one of the sheds in our barn and later in the woods. But, having more buildings, we then converted the goose house into a chicken house and had about eighteen chickens of many colors!

On February 27, 2007, our lovely granddaughter Melissa, known as Milly, was born, sister to Taia. She was a huge character from the beginning. Eighteen months later Nim produced Tom on August 14, 2008, to complete our trio of fantastic grandchildren.

When Sally Thompson was over here from South Africa for Nim's wedding, we visited Sissinghurst Castle Gardens—created by Vita Sackville-West, poet and writer, and her husband Harold Nicholson, author and diplomat. They are amongst the most famous gardens in England and were divided into a series of "rooms." We so enjoyed the visit that friends recommended we also visit Hidcote Manor Gardens near Chipping Norton, where Major Lawrence Johnston created one of the best-known and most influential arts and crafts gardens in Britain, with its linked "rooms" of hedges, rare trees, shrubs and herbaceous borders.

We were inspired and decided to create our own "Stitchcote." Being a structural engineer and having spent most of my life working with architects, I knew I could appreciate great design—but I lacked the skill to be artistically creative. I therefore decided we needed to find a good landscape gardener to help us. Having examined a number of books on gardens, I found a book by Penelope Hobhouse which I really liked, with mown paths through paddocks of wildflowers and interesting hedges of hornbeam like Hidcote.

I found her telephone number and rang her up. I explained what I wanted to do and asked her to come for lunch one Sunday. She at first accepted but then

Grove House: Lime Walk with water fountain and three graces in distance (author's collection).

rang back and said her partners felt we might do something which could damage her name. I then unashamedly name dropped and said I completely understood because I was an engineer who worked with Sir Norman Foster and Sir Richard Rogers and that my partner was Philip Dowson, President of the Royal Academy. I assured her that we would behave! Anyway, she agreed to come to Stitchcombe, walk around, give me the benefit of some ideas and have lunch—all for a fee of £200.00. Her first question was, "How many gardeners do you have?" Upon my answer, she said, "It will obviously have to be low maintenance!"

After saying she wasn't really any good at this sort of thing, Penelope eventually said that I should imagine planting the whole paddock next to the house with trees but cut out a square matching and alongside our house and walled garden. Then enter a "secret garden" through the granary (formerly the vineyard shop)—then enter a lime walk with a circular pool in the middle, and then the four crisscross avenues would radiate off the pool all lined with hornbeam hedges. I could build the lake I wanted by the road and then plant wildflowers in the enclosed triangular paddocks. This was exactly what we wanted. It was only afterwards that I learned that she had just been appointed by the Queen Mother to design a garden for her at Hampton Court.

We duly laid out the garden as Penelope had proposed and placed our 200-year-old statue of "the three graces" at the east end of the Lime Walk. And we built a gazebo at the top end of one of the hornbeam avenues. We also planted approximately 7,000 English broadleaf trees.

By 2012 I had become fed up with letting our fields and stables to horse ladies because over the years, the horses were damaging the fields and eating the fences and stables. I asked an old friend, Sid, the ex-policeman turned gamekeeper, for advice, and he suggested that we git rid of the horses and go in for sheep!

In 2013 our sheep phase started. I bought six four-month-old Hampshire Down ewe lambs from the current president of the Hampshire Down Sheep Breeders Association. The following year we bought our first Hampshire

Grove House: Nigel and Nicky on terrace with statue by Helen Sinclair (girl with the mountain flowers) (author's collection).

Grove House: Florence, circa 1910, picking a pear from the pear tree (author's collection).

ram so that at Christmas time 2014–15 we had our first lambs. Over the next six years until 2020 we had 74 lambs. Every year we managed to sell nearly all rams for breeding purposes and only a few sheep for the food chain.

The grandchildren, and particularly Tom, helped with the sheep. They loved all the lambs and driving around in my Kawasaki Mule—4-wheel drive. In fact, Tom became very accomplished at handling the sheep, even the large rams—he had no fear! I am sure the two girls would have liked to help more but they lived too far away in the South of France.

The Hampshires are a most attractive-looking sheep. Having beautiful grass in the avenues, wildflowers in the meadows and these good-looking sheep in our fields gives us much pleasure as we walk around and enjoy our estate.

In fact, having the dogs, chickens, geese, piglets, lambs and bees (two or three hives over the past ten years) has been a great attraction for the children and grandchildren to come and stay at Grove. Every Christmas they all come to stay—the grandchildren insist! Our family Christmases are huge fun, and they have

Nigel with Tom and sheep (author's collection).

From left, Taia, Tom and Milly with three Hampshire Down lambs (author's collection).

41. Grove House and Estate

improved over the years, particularly as Taia has become a very accomplished piano player, singer and composer and writes her own songs. Both Milly and Tom also play. So we have entertainment around the piano as well as playing traditional games such as charades.

Not so long ago, Milly, Sam's second daughter, asked me what will happen to Grove when Nicky and I die. I said it would go to the family, and she quickly said Tom and Nim could have the cottages and they would have the house—because Sam was our only son and the oldest son should have the first choice. I said, "No, I'm afraid not, because Nim is the eldest, and anyway, Grove will be divided between Nim, Sam and Phoebe."

Nicky and I very much love Grove House and our estate. For our fiftieth wedding anniversary we gave each other another statue by Helen Sinclair—a girl holding flowers. We have her standing on the terrace picking a rose. We have named her Florence as we found in the house a picture of Florence Chambers who lived at Grove between around 1880 until about 1939. She was picking a pear from the tree at about that same spot probably around 1910!

For my eightieth birthday my whole family gathered plus one of Peter's

The family in 1980. From left: (top) Nigel Thompson, not named, Phoebe Thompson, Oliver Cohen, Nim Thompson, Nicky Thompson, Noo Weber, Emma Reeves, Bridget Bonnett, Daphne Corke, Elizabeth Cohen, Edwina Cohen, Charlotte Cohen and George Cohen (foreground) not named, Fleur Weber, Simon Weber and Sam with Tosca the dog (author's collection).

daughters, Jude, John, her husband and two children Kate and Harry, plus Peter's other grandson Luke, Frances' first son and his girlfriend Grace, plus Simon, Nicky's nephew, with his wife Pip and two children Monty and Lexie. All came down to Grove House, and we had lunch at the Bell Inn in Ramsbury to celebrate a great day.

The family on my eightieth birthday in 2019. From left: (top) Andrew Harris, Nim Harris, Nigel Thompson, Tonia Thompson, Taia Thompson, Phoebe Thompson, Nicky Thompson, Milly Thompson, Sam Thompson, Harry Wixley, Simon Weber, Pip Weber, Jude Wixley, John Wixley and Kate Wixley (foreground) Grace Di Benedetto, Luke Harrison, Monty Weber, Tom Harris holding Cleo and Lexie Weber with Doody the dog (author's collection).

Appendix I
Nigel Thompson Selected Titles and Awards

Sir Nigel Cooper Thompson KCMG, CBE, C.Eng., FICE, FIStructE

May 2002—By the Command of Our Sovereign Queen Elizabeth the Second Nigel Cooper Thompson Esquire Commander of Our Most Excellent Order of the British Empire be appointed Knight Commander of Our Most Distinguished Order of Saint Michael and Saint George (KCMG).

Other Selected Titles and Awards

July 15, 2003—Sir Nigel Thompson KCMG, CBE, being also a Freeman of The City of London, was admitted to the Livery of the Worshipful Company of Engineers.

February 12, 2003—Sir Nigel Cooper Thompson KCMG, CBE, Citizen and Engineer of London was admitted into The Freedom of The City of London in the 52nd year of the reign of Queen Elizabeth the Second. (Author's note: one of the major perks of being a Freeman of the City of London is that you are allowed the privilege of driving your sheep over London Bridge into the City!)

2001—Nigel Thompson CBE, Deputy Chairman, Ove Arup and Partners was awarded The Association of Consulting Engineers (ACE) Outstanding Contribution Award in recognition of outstanding contribution to the consulting engineering profession.

December 20, 2000—By Her Majesty Queen Elizabeth the Second's Command Nigel Cooper Thompson was appointed to the rank of Major in the Royal Engineer's Logistic Staff Corps.

June 13, 2000—Nigel Cooper Thompson, being a Chartered Engineer, was transferred from being Member to the class of Fellow of the Institution of Civil Engineers (FICE).

June 15, 1996—By the Command of Our Sovereign Queen Elizabeth the Second Nigel Cooper Thompson Esquire was appointed to be a Commander of the Civil Division of Our said Most Excellent Order of the British Empire (CBE).

January 2, 1992—Nigel Cooper Thompson was transferred from Member and elected Fellow of the Institution of Structural Engineers (FIStructE).

February 1960—Nigel Cooper Thompson was awarded a scholarship by the Federation of British Industries (FBI) and the British Council (BC) for two years post graduate training in the United Kingdom with Ove Arup and Partners in London. The following year the Federation of British Industries became the Confederation of British Industries (CBI)

Appendix II
Terms and Abbreviations

Terms

Air Rights Buildings—Air rights are the property interest in the "space" above the Earth's surface. Generally, owning or renting land or a building includes the right to use and develop the space above the land without interference by others.

Crown Agents—An international development company with head office in the United Kingdom. Its main focus is to help governments around the world to increase prosperity, reduce poverty and improve health by providing consultancy, supply chain, financial services and training.

European Bank for Reconstruction and Development (EBRD)—Uses investment as a tool to build market economies across the former Eastern Bloc and in more than 30 countries from Central Europe to Central Asia. Headquartered in London, the EBRD is owned by 69 countries and two EU institutions, the sixty-ninth being India since July 2018. Despite its public sector shareholders, it invests in private enterprises, together with commercial partners.

Listed Buildings in the United Kingdom—Grade I listed buildings are buildings of exceptional national interest (only about 2 percent of all listed buildings are in this category); Grade II listed buildings are particularly important buildings of more than special interest (about 4 percent of all listed building are in this category); Grade II listed buildings are buildings of special interest which warrant every effort being made to preserve them. Listed buildings include all buildings built before 1700 which survive in anything like their original form.

London Club—An informal group of private creditors on the international stage, similar to the Paris Club of public lenders.

Loo—Comfort station, lavatory, toilet, bathroom, washroom, restroom, water closet (wc).

MI6—An organization formed to protect the United Kingdom's people, economy and interests from external threats.

Old Pauline—Former pupil of St. Paul's School, West London.

Paris Club—An informal group of private creditors on the international stage, similar to the London Club of public lenders.

Saints—Inhabitants of the island of St. Helena.

World Bank—An international financial institution that provides loans and grants to the governments of low- and middle-income countries for the purpose of pursuing capital projects. It comprises two institutions: the International Bank for Reconstruction and Development (IBRD) and the International Development Association (IDA).

Abbreviations

ABK—Ahrends, Burton and Koralek

AFFORD—African Foundation for Development

AMICE—Associate Member of the Institution of Civil Engineers

AMIStructE—Associate Member of the Institution of Structural Engineers

ARK—Action for the River Kennet

ARP—Air Raid Precautions

BAC—Business Advisory Council

BAC for SEE—Business Advisory Council for South Eastern Europe

BAC to the Stability Pact—Business Advisory Council to the Stability Pact (for South Eastern Europe)

BAG—British Airports Group

BCB—British Consultants Bureau

BCO—British Council for Offices

BDU—Business Development Unit

BMS—Building Management System

BOTB—British Overseas Trade Board

BR—British Rail

BRITFOR—British Armed Forces

BT—British Telecommunications

BTI—Binding Tariff Information

BUILD—Building Understanding through International Links for Development

CAA—Civil Aviation Authority

CARDS—Community Assistance for Reconstruction, Development and Stabilisation

CBI—Confederation of British Industry

CCF—Combined Cadet Force

CEMVO—Council for Ethnic Minority Voluntary Organisations

CFD—Computational Fluid Dynamics

CIA—Central Intelligence Agency

CLASP—Consortium of Local Authorities Special Programme

CPRE—Campaign to Protect Rural England (CPRE The countryside charity)

CRISPR—Clustered Regularly Interspaced Short Palindromic Repeats

DDT—Dichlorodiphenyltrichloroethane

DETR—Department of the Environment, Transport and the Regions

DFID—Department for International Development

DNA—Deoxyribonucleic acid

DoT—Department of Transport

DTI—Department of Trade and Industry

EBRD—European Bank for Reconstruction and Development

EC—European Commission

ERTP—Economic Recovery and Transition Programme

EU—European Union

FBI—Federation of British Industries

FCO—Foreign & Commonwealth Office

FRY—Socialist Federal Republic of Yugoslavia

HIV—Human Immunodeficiency Virus

HMG—Her Majesty's Government

IBM—International Business Machines Corporation

ICW—International Community of Women with HIV/AIDS

ILP—Institution of Lighting Professionals

IMF—International Monetary Fund

ISG—Infrastructure Steering Group

IT—Information Technology

KEK—Korporata Energjetike e Kosovës

KFOR—Kosovo Force

KISR—Kuwait Institute for Scientific Research

LCCI—London Chamber of Commerce and Industry

LDW—Llewelyn-Davies Weeks

M&E—Mechanical & Electrical (Engineering)

MEP—Mechanical, Electrical and Plumbing (Engineering)

MoD—Ministry of Defence

MOU—Memorandum of Understanding

MTB—Motor Torpedo Boat

NATO—North Atlantic Treaty Organization

NURRU—Network of Ugandan Researchers & Research Users

OAP—Ove Arup Partnership

OP—Old Pauline

OSCE—Organisation for Security and Cooperation in Europe

PC—Parish Council

PCC—Parochial Church Council

PPE—Personal Protective Equipment

PPG—Planning Practice Guidance

PPP—Public-private Partnership

RE—Royal Engineers

RHWL—Renton Howard Wood Levin Architects

RIBA—Royal Institute of British Architects

RMS—Royal Mail Ship

SCSP—Special Coordinator of the Stability Pact

SECI—Southeast European Cooperative Initiative

SEE—South Eastern Europe

SHCA—Swanke Hayden Connell Architects

SHCPLC—Saint Helena Corporations PLC

SHELCO—St. Helena Leisure Corporation

SOM—Skidmore Owings & Merrill

TAFKO—Task Force for the Reconstruction of Kosovo (EC)

UCK—Kosovo Liberation Army

UDI—Unilateral Declaration of Independence

UKOWLA—United Kingdom One World Linking Association

UNDP—United Nations Development Programme

UNMIK—United Nations Interim Administration Mission in Kosovo

UNSCR—United Nations Security Council Resolution

UPS—Uninterruptible Power Supply

VE—Victory in Europe

VSO—Voluntary Service Overseas

WTO—World Trade Organization

Chapter Notes

Chapter 1

1. In November 1938, UK Prime Minister Neville Chamberlain placed Sir John Anderson in charge of Air Raid Precautions, in anticipation of German bombing attacks. Anderson immediately commissioned an engineer, William Patterson, to design a small and cheap bomb shelter that could be erected quickly and simply in people's gardens. The first "Anderson" shelter was erected in a garden in Islington, London, on February 25, 1939. Anderson shelters were designed for six people. The main part of the shelter was formed from six corrugated steel panels. Flat, corrugated steel panels were bolted on to form the sides and end panels, one of which contained the door. Each shelter measured 4.6 ft (1.4 m) wide x 6.6 ft (2 m) long by 6 ft (1.8 m) tall. Once constructed, the Anderson shelter was buried more than 3.3 ft (1 m) in the ground and then covered over with a thick layer of soil and turf. Anderson shelters were free to those with an annual income of less than £250 per annum (U.S.$1,220 at 1938 exchange rates). For those who didn't fall into this category, the price was £7 (U.S.$34 at 1938 exchange rates). Approximately 3.5 million Anderson shelters were built either before the war had started or during the war.

2. The Vickers anti-aircraft guns that Britain used to target Germany's Luftwaffe bombers were known as ack-ack guns because of their noise. Some 10,000 were built, in static and mobile versions, between 1937 and 1945. Each weighed 20,541 lb (9,317 kg), and the fixed version was bolted to the ground in cement (the gun pit wall was four foot down because, underneath, there were bunks to sleep on in a lull). Overall length was 28 ft 3 in (8.6 m), barrel length 15 ft 5 in (4.7 m), width 7 ft 10 in (2.4 m) and height 8 ft 2 in (2.5 m). Caliber was 3.7 in (94 mm) and shell weight 28 pounds (13 kg). Maximum firing range was 3.5 miles (5.6 km) horizontal, 7.5 miles (12 km) inclined, and up to 45,000 ft (13.7 km) ceiling (on the Mark VI version). Each gun required up to 11 men to operate it. One man stood on the platform, two men turned the handles and the others loaded the shells.

3. Anderson shelters were very effective at saving lives and preventing major injuries during air raids, but they were very cold during the winter months. For winter use, the government developed the "Table (Morrison) Indoor Shelter." The Morrison shelter was designed by John Baker and named after the minister of home security, Herbert Morrison. It was supplied in kit form for assembly (bolting together) in the home. Each Morrison shelter comprised more than 350 parts, but the principal elements of construction were a steel top (like a tabletop) and wire mesh sides (one of which could be lifted open to act as the door). People would sleep under the 'tabletop' at night and would use the shelter as a real table for the rest of the time. The Morrison shelter was not designed to survive a direct hit from a bomb but was effective at protecting people from the effects of bomb blast. More than 500,000 Morrison shelters were made, and they were given free of charge to families earning less than £350 (U.S.$1,700) a year.

4. The V-1 flying bomb, also known to the Allies as the "buzz bomb" or "doodlebug," was an early cruise missile and the only production aircraft to use a pulsejet for power. It was developed by the Nazi German Luftwaffe under the codename "Cherry Stone" at Peenemünde Army Research Center in 1939. The V-1 was the first of the "Vengeance weapons" series (V-weapons or Vergeltungswaffen) deployed for the terror bombing of London. Because of its limited range, the thousands of V-1 missiles launched into England were fired from launch facilities along the French (Pas-de-Calais) and Dutch coasts. The Wehrmacht first launched the V-1s against London on June 13, 1944, one week after—and prompted by—the successful Allied landings in France. At peak, more than 100 V-1s per day were fired at South-east England. There were 9,521 in total, decreasing in number as sites were overrun until October 1944, when the last V-1 site in range of Britain was overrun by Allied forces.

Chapter 2

1. Willington School was originally located at No. 3 Dealtry Road, Putney. In 1889 it moved to Willington House in Colinate Road, just off the Upper Richmond Road, Putney (the address that I attended), and the school is now located on the Worcester Road, Wimbledon.

2. Building societies exist in the United Kingdom, Australia and New Zealand. A building society is a financial institution owned by its members as a mutual organization. It offers banking and related financial services, especially savings and home mortgage lending to members. The term "building society" was first used in the nineteenth century in Great Britain. The Norwich Building Society was founded in 1852 under the title of the Norwich and District Provident Permanent Benefit Building and Freehold Land Society.

3. My relatives sold small items used in sewing such as buttons, zips and thread through to material for sofas, curtains, gentlemen's suits etc. The word *haberdasher* has a different meaning in North America, where it indicates a dealer in men's clothing.

4. St. Paul's School takes its name from St Paul's Cathedral in London, where a cathedral school had existed since around 1103. By the sixteenth Century, it had declined and, in 1509, a new school was founded by John Colet, Dean of St Paul's Cathedral, on a plot of land to the north of the Cathedral. This was destroyed with the cathedral in the Great Fire of London in 1666. The school was rebuilt in 1670 and again, in Cheapside, in 1822. Towards the end of the nineteenth century, because London was expanding and its residents were moving away from the city, it was decided that the school should move to larger premises. It is this new, bigger building, designed by the architect Alfred Waterhouse and dating from 1884, that I attended at Hammersmith. St. Paul's School is today located on a 43 acre (17.4 hectare) site by the River Thames in Barnes, London.

5. An Old Pauline is an alumni of St. Paul's School. Famous Old Paulines include John Milton (1608–1674), author of the epic poem *Paradise Lost*, Samuel Pepys (1633–1703) and the prolific architect George Dance the Younger (1741–1825). In the context of this book, the most significant Old Pauline is the architect/engineer David Asher Alexander (1768–1846) who was the surveyor to the London Dock Company (1796–1831) and was responsible for all the buildings at the London Docks during that time.

6. *Loo* is a British informal noun equivalent in meaning to toilet, bathroom or restroom.

7. Horlicks is a sweet, malted-milk, hot-drink powder developed by company founders James and William Horlick in 1873. It was first sold as "Horlick's Infant and Invalids Food" and later, in the early twentieth century, as a powdered meal replacement drink mix. It was then marketed as a nutritional supplement and manufactured by GlaxoSmithKline (Consumer Healthcare) in the United Kingdom, Malaysia, Australia, New Zealand, Hong Kong, Bangladesh, India, Sri Lanka and Jamaica. Horlicks is now produced by the Anglo-Dutch company Unilever. (Wikipedia).

8. The Combined Cadet Force (CCF) is a youth organization founded in 1948 in the United Kingdom and sponsored by the Ministry of Defence (MOD). It operates in schools and normally includes Army, Royal Navy and Royal Air Force sections. Its aim is to "provide a disciplined organisation in a school so that pupils may develop powers of leadership by means of training to promote the qualities of responsibility, self-reliance, resourcefulness, endurance and perseverance." (Wikipedia).

9. The Scouting movement dates back to the 1908 publication of *Scouting for Boys* written by Robert Baden-Powell. Its aim is to contribute to the development of young people in achieving their full physical, intellectual, emotional, social and spiritual potentials as individuals, as responsible citizens and as members of their local, national and international communities.

Chapter 3

1. Midland Bank Plc. was one of the Big Four banking groups in the United Kingdom for most of the twentieth century. By 1934, it was the largest deposit bank in the world. In 1992, it was taken over by HSBC Holdings plc. which phased out the Midland Bank name by June 1999 in favor of HSBC Bank.

2. "Lloyd's of London, generally known simply as Lloyd's, is an insurance and reinsurance market located in London, United Kingdom" (Wikipedia). Its roots are in marine insurance, and it was founded by Edward Lloyd at his coffee house on Tower Street around 1686.

3. "The Union-Castle Line was a British shipping line that operated a fleet of passenger liners and cargo ships between Europe and Africa from 1900 to 1977" (Wikipedia).

4. Deck tennis, a hybrid between tennis and quoits, is played with either a rubber disk or ring, or a similarly sized rope ring. It is one of many sports and leisure activities devised to pass the time on the long ocean voyages that predated the era of popular, affordable, passenger international air travel.

5. The Kariba Dam is a double curvature concrete arch dam in the Kariba Gorge of the Zambezi River basin between Zambia and Zimbabwe. The dam stands 128 m (420 ft) tall and 579 m (1,900 ft) long. The dam forms Lake Kariba which extends for 280 km (170 miles) and holds 185 cubic kilometers (150,000,000 acre-ft.) of water.

6. Impregilo was founded in 1959. It became Salini Impregilo SpA in 2014 and is now, since

2020, part of WeBuild, an Italian industrial group headquartered in Milan and specializing in construction and civil engineering. WeBuild operates in more than 50 countries across all continents and has 35,000 employees.

Chapter 4

1. Peter Dunican (1918–1989) was inspirational to both my professional and personal life. Peter joined Arup & Arup Ltd. in 1943, when it was a firm of contractors and designers of civil engineering works. He moved with Ove Arup when Ove left Arup & Arup to set up his independent consulting engineering practice. Ove was the philosopher and visionary of the engineer as a creative collaborator in the design of buildings (rather than simply an analyst and number-cruncher). But it was Peter Dunican who made it work so that, between them, they raised the aspirations and the profile of the building engineer. Peter was attracted especially to "socially useful" building—housing, schools, universities and hospitals. He wanted to design industrialized housing that would quickly meet the needs of a nation whose housing stock had been devastated by war. Peter would become a part-time director of the National Building Agency set up by the Labour Government in 1964 "...to help the industry to deal with the great upsurge of demand now facing it...."

2. Interestingly, some of the design principles demonstrated at Northwick Park derive from hospital design by Isambard Kingdom Brunel during the Crimean War of the 1850s. In 1855 British politicians and public alike were outraged that 42 percent of their soldiers were dying not on the field of battle but during the subsequent care of their wounds. The role of Florence Nightingale in improving this situation is universally acknowledged. But Brunel's innovation in hospital design working with Florence Nightingale is often overlooked because of his huge achievements in other fields of engineering.

Essentially, the Government of the 1850s wanted a hospital building that could be prefabricated for shipping piece-small from the UK to the Dardanelles. This building would be of repetitive structure, for ease of assembly and future extendibility. Airflow through the building would reduce the cross-contamination that was the cause of many unnecessary patient deaths from infection. Brunel arrived at a steel and timber-framed solution within days, organized fabrication of the components and arranged the trial erection of a prototype on the premises of the Great Western Railway at Paddington in west London. Within a matter of weeks, during 1855, Brunel transformed a blank piece of paper into a fully erected hospital at Renkoi in the Dardanelles.

The design principles developed by Brunel were taken up by the UK's Nuffield Health Foundation in its 1950s' research into the future of hospital design. In 1958 Lord Llewelyn-Davies (1912–1981), Director of the Division of Architectural Studies of the Nuffield Foundation, and John Weeks, the Deputy Director, began a process that would see these principles widely implemented by forming the practice of Llewelyn-Davies Weeks Forestier-Walker & Bor in 1959. In 1962 this practice and Arup were commissioned to work together on Northwick Park Hospital, Harrow, a 600-bed district general hospital and 200-bed national clinical research centre, together with residential accommodation and ancillary buildings. Northwick Park embodied the Llewelyn-Davies and John Weeks 'loose-fit' concept of a building envelope that could be altered and extended at will, readily accommodating growth and the replacement or upgrading of building services.

3. John Weeks (1921–2005) was an architect and health services planner who pioneered a new kind of architecture capable of responding to the complex and continually changing needs of large institutions, including hospitals, universities and research organizations. His design approach, "indeterminacy," allows a building to grow and change with the growth and change of the institution it houses. John designed each building with a "backbone" of routes for circulation and services, fixed for the life of the building. On to these routes he attached all the independent elements of the building so that each element is endowed with an individual capacity for growth and change. He carried this philosophy into the detailed designs, down to individual laboratories or consulting rooms. His approach did not entail increased building and engineering costs and usually resulted in more economical planning solutions than traditional designs.

4. Llewelyn-Davies Weeks (originally Llewelyn-Davies Weeks Forestier-Walker & Bor) had an early mission statement: "Increasing concern with the development and management of health services is shared by communities throughout the world. To meet the varied needs of clients in this field, we have developed a comprehensive array of services—ranging from strategic planning at the national or regional level, through the design and commissioning of new hospitals, to management consultancy within existing institutions."

Chapter 5

1. My travel companions to Turkey, Dux and Monique Schneider, were both authors who wrote many books, including several on Turkey. Monique wrote under the name Frances Oliver, and among her 13 books, all published by Perron Press, Penzance, she wrote *Jaundiced in Antalya* which includes the six weeks I spent with them

climbing mountains and visiting Lake Van and the Armenian Church of Akhtamar. In fact, I have taken a sentence or two directly from her book!

2. Sainte Marie de La Tourette is a Dominican Order priory located on a hillside near Lyon, France. It was the final building (completed 1961) of architect Le Corbusier (1887–1965). The irregularly sloping site allowed the architect to explore a unique concept—the upside-down city. Le Corbusier raised the structure on piers, allowing the terrain to undulate, and provided circulation at the top of the structure.

3. Completed in 1954 in the commune of Ronchamp, in eastern France, sits one of Le Corbusier's most unusual buildings, the chapel of Notre Dame du Haut (Our Lady of the Heights). Appropriately, the most striking part of Ronchamp is the curved roof which peels up towards the heavens, appearing to float above the building (it is supported by embedded columns in the walls, which create a gap between the roof and walls for a slither of clerestory light).

Chapter 6

1. Christine Keeler (1942–2017) was the most photographed face of 1963. In 1961, aged 19, she had an affair with the UK secretary of state for war, John Profumo, at the same time as she was in a sexual relationship with a Russian spy, Eugene Ivanov. The possibility of "pillow-talk" was obvious. The political time-bomb exploded into a court case in 1963 against Keeler's friend and mentor, the high-society osteopath Stephen Ward, who committed suicide before the "guilty" verdict was handed down. Keeler and her friend Mandy Rice-Davies had visited America in what Keeler described as a "daft summer" of 1962. The Kennedy White House took a keen interest in Keeler's affairs, which were documented by J. Edgar Hoover in secret FBI papers filed under the code name *Bowtie*. It amused Keeler to find out, many years later, that the U.S. attorney general Robert F. Kennedy had been readily aware of who Hoover meant by "Christine" but had to be supplied with her surname.

2. The Mini is a small economy car manufactured by the English-based British Motor Corporation (BMC) and its successors from 1959 to 2000. The original is considered an icon of 1960s British popular culture, and its design influenced all carmakers. My Mark I Mini was marketed in North America as the Austin 850. (Christine Keeler owned a silver Mark I Mini at the height of the Profumo scandal.)

3. The Trout Inn (often simply referred to as "The Trout") features in Evelyn Waugh's novel *Brideshead Revisited* and in Colin Dexter's Inspector Morse detective books and TV series, which were written and filmed in and around Oxford.

Chapter 7

1. Greek Cypriot attacks on the Turkish Cypriot community had led, on March 4, 1964, to the United Nations Security Council adopting a resolution (186), which provided for stationing a United Nations Peace Keeping Force in Cyprus (UNFICYP). But the arrival of UNFICYP did not bring an end to the violence, which continued to disrupt civilian life on the island.

Chapter 8

1. "Buckler's Hard is a hamlet on the banks of the Beaulieu River in the English county of Hampshire" (Wikipedia). It is part of the 9000-acre (3600 hectare) Beaulieu Estate.

Chapter 9

1. The Everyman, in Heath Street, Hampstead, London, opened as a cinema on December 26, 1933. Sir Gerald du Maurier (1873–1934), the famous actor/manager, and resident of Hampstead, presided at the opening of the Everyman and the first programme consisted of *Le Million*, *Turbulent Timber*, a Mack Sennett comedy, a Disney cartoon and Paramount News.

2. "*L'Année dernière à Marienbad* (released in the U.S. as *Last Year at Marienbad* and in the UK as *Last Year in Marienbad*) is a 1961 French-Italian Left Bank film directed by Alain Resnais from a screenplay by Alain Robbe-Grillet. Set in a palace in a park that has been converted into a luxury hotel, it stars Delphine Seyrig and Giorgio Albertazzi as a woman and a man who may have met the year before and may have contemplated or started an affair, with Sacha Pitoëff as a second man who may be the woman's husband. The characters are unnamed. The film is famous for its enigmatic narrative structure, in which time and space are fluid, with no certainty over what is happening to the characters, what they are remembering, and what they are imagining." (Wikipedia).

Chapter 10

1. "A bistro is, in its original Parisian incarnation, a small restaurant, serving moderately priced, simple meals in a modest setting with alcohol. Bistros are defined mostly by the foods they serve. French home-style cooking is typical." (Wikipedia).

Chapter 11

1. At the major, new, 600-bed York District Hospital I worked with Ernie Freeman as my

project engineer and with Nick Taylor as site engineer. York has a 'zoned' plan compared with the 'differentiated' plan of Northwick Park. Both hospitals have reinforced concrete frame structures. At York we further advanced the art of hospital building design by developing an innovative, highly penetrable but flat floor system. York would be one of the most economic district general hospitals ever built. Because it would have large areas of slab not covered with false ceilings, we looked at forming the coffer with a permanent lightweight concrete cellular block. The resultant plate floor had all the advantages of a coffered floor with regard to perforation and inherent stiffness. It also had the advantage of providing a flat soffit ready for decoration and to receive partitions on a smaller superimposed grid. In this case the planning module was 300 mm and the primary structural module 900 mm, with the vertical structure placed on a tartan grid generally at 7.2 m centers in both directions. This design was produced to accord in principle with the publications on dimensional coordination published by the then-Ministry of Health and Ministry of Public Buildings and Works.

York was a result of the line of hospital development that began with Nuffield and was developed through Northwick Park. There was a second line of hospital development in the 1960s that began with Greenwich District General Hospital, which introduced the interstitial floor concept to the UK. Greenwich led to 'Best Buy' hospitals, which took a similar ring plan to Greenwich but eliminated the need for air-conditioning by inserting courtyards. Best Buy was indeed economical—similar to York DGH in cost terms—but Best Buy didn't cater for future phasing, so growth was not really feasible, and even potential for change-in-use was severely limited.

However, out of Best Buy came The Harness Hospital Programme, with Arup being commissioned as project consulting engineers on the first prototype Harness hospital, to be built at Shoreham in Sussex. We were also engaged separately as subsystem designers for one of the cladding systems that the individual client for a Harness hospital could choose. At a later stage we were asked to do a research project with Syska and Hennessy from New York and to report on the system as it was developed. I was lucky to persuade Martin Manning, a very bright young structural engineer, to join me for this project. Economic restraints and our research work led to the demise of Harness, but the progeny of Harness–Nucleus retained some of the Harness concepts of medical and room planning. Meanwhile, the line of hospital design developed at Northwick Park, and refined at York, would be applied and further refined in the UK (e.g., St Mary's, Paddington), in Australia (Westmead Teaching Hospital, New South Wales) and in Iran.

2. The National Hospital for Nervous Diseases in London is now known as the National Hospital for Neurology and Neurosurgery (NHNN) and is the UK's largest dedicated neurological and neurosurgical hospital. It provides comprehensive services for the diagnosis, treatment and care of all conditions that affect the brain, spinal cord, peripheral nervous system and muscles.

3. The architectural practice Levitt Bernstein was founded in 1968, shortly after David Levitt met David Bernstein whilst working on the groundbreaking Brunswick Centre, now a Grade II listed residential and shopping center in Bloomsbury, London. They decided to set up a practice focused on revitalizing London's decaying inner city. The capital was also in the throes of a housing crisis, and the pair were dedicated to creating better homes for all.

4. Now Clarion Housing Group, the UK's largest housing association.

Chapter 12

1. Frei Paul Otto (1925–2015) was renowned for designing lightweight tensile and membrane structures, including the roof of the Olympic Stadium in Munich for the 1972 Summer Games.

2. Rolf Gutbrod (1910–1999) collaborated with Frei Otto and Arup on the hotel and conference center in Mecca, Saudi Arabia, 1966–74.

3. In Iran, Reza Shah Pahlavi, the founder of the Pahlavi dynasty, had ruled from 1925–1941. He wanted Tehran to be considered the equal of cities like Ankara, Paris and London. Tehran grew from a modest town concentrated around the bazaar and Golestan Palace to a city of more than 4.5 million people. Engineers and architects cut through the heart of the sprawling city and created a network of traffic arteries, parks and administrative structures. In 1941 Reza Shah was succeeded by his son Mohammad Reza, who was an advocate of the *Great Civilization*. Tehran became known to westerners as the location of the 1943 meeting of Joseph Stalin, Winston Churchill and Franklin D. Roosevelt. The three world leaders assured Iran's new ruler of his country's independence and promised that they would compensate Iran for its war efforts. The latter promise was not kept, but Tehran continued to prosper anyway. In 1963 Mohammad Reza launched the White Revolution, a broad attack on the ills of Iranian society that focused on land reform. He started building highways, palaces, massive dams and a few model cities.

4. Mohammad Reza Shah Pahlavi (1919–1980) was the Shah (King) of Iran from 1941 until his overthrow in the Iranian Revolution of February 11, 1979. Because he was the last Shah, he is often simply referred to as *the Shah*.

5. Christopher Chataway (1931–2014) was an Olympic athlete best remembered for, with Chris Brasher, pacemaking Roger Bannister in the world's first sub-four-minute mile at Oxford in 1954. Chataway became a TV news broadcaster

and Conservative member of parliament before retiring from politics in 1974 (aged 43) and beginning a career in business as managing director of Orion Bank (which was later acquired by the Royal Bank of Canada).

6. The Shahyad (Memory of the Shah) Monument at Azadi Square was built in 1971–1972. It houses a museum in its basement and exhibition halls on its upper floors. Principally, however, it serves as the gateway to bustling Tehran, symbolizing Iran's glorious past and promising future through its fusion of traditional Islamic and modern Iranian architecture. The unique 190 ft (58 m)-tall structure, with four massive columns forming four arches, was designed by an Arup team which included the firm's future chairman Duncan Michael. The architectural design of the cut-marble Shahyad had been worked up to mirror the 13,770 ft (4,197 m) tall Mount Damavand, the tallest peak in the mountain range just north of Tehran.

7. Ruhollah Khomeini (1902–1989) was the supreme leader of Iran, founding the Islamic Republic of Iran after leading the 1979 Iranian Revolution which ended the 2,500 year old Persian monarchy.

8. Michael ("Mick") Lewis (1927–2011) was a civil engineer who, with Jack Zunz, opened Arup's first office in Africa in 1954. They led the firm's operations in southern Africa until 1962. Michael Lewis went to Sydney in 1963 to coordinate the design, management and construction of the Sydney Opera House and was a founder of Ove Arup & Partners Australia. On his return to London in 1974, he became responsible for all of the firm's civil engineering activities.

9. Jaquelin ("Jaque") T. Robertson (1933–2020) was an American architect and urban designer. He was descended from U.S. fourth President James Madison, Jr. (1751–1836) and U.S. twelfth President Zachary Taylor (1784–1850). Robertson worked in New York City planning, founded the New York City Urban Design Group, was the first director of the mayor's office of Midtown Planning and Development and was a city planning commissioner. In 1975 he went to Tehran to direct the planning and design of the new capitol center Shahestan Pahlavi in the Abbas Abad district of the city.

10. Queen Noor, *Leap of Faith: Memoirs of an Unexpected Life*, Weidenfeld & Nicolson, 2003, 36–37.

11. Martin Manning ultimately became Arup's global structural engineering leader—I was lucky to have him with me in Iran!

12. Holger Koch-Nielsen later wrote a book, *Stay Cool: A Design Guide for the Built Environment in Hot Climates*, which advises on ways of reducing the need for, and dependence on, technology for buildings in hot climates.

13. Alvar Aalto (1898–1976) was a Finnish architect who had been a friend of Sir Ove Arup's for many years. Aalto was considered so important that his national airline, Finnair, would delay takeoff if he was expected for a flight but was late turning up. He apparently enjoyed his "grand entrances" so much that, if he arrived on time, he'd instruct his chauffeur to drive around Helsinki Airport until he was late for boarding.

14. Iran Museum of Modern Art in southern Iran, close to the ancient capital of Persepolis, was Alvar Aalto's final building design. The site was atop a prominent hill affording views of the whole of the city of Shiraz which, at the time, had a population of about 300,000. On top of the hill a small reservoir was kept pumped full. This water supply was used to help cultivate the trees and shrubs that would, in time, be used to landscape the sculpture gardens and walks that would surround the museum buildings. Looking from the city, the site had a backcloth of arid hills. To preserve this outlook, Aalto designed a two-story building in the segmented form of an outcropping, with main galleries radiating to a maximum height of about 59 ft (18 m) above normal ground level. His large areas of blank walling would be clad in a dark-red brick that would contrast with the surrounding pale desert tones. Almost 32,300 ft^2 (3,000 m^2) of exhibition spaces would be complemented by lecture halls, a library, cafeteria and administration facilities. The open but divisible exhibit floor would gather its illumination directly from the sky, through glazed and louvered roof surfaces. Aalto identified three aspects of the Iranian sun: glare, heat and light. With his design, he said, "I am stopping the glare, barring the heat, but capturing the light for my building so that you can still see the blue light of Shiraz." Structural work could not start until a level platform had been created, and this necessitated the removal of up to 33 ft (10 m) of rock from the top of the hill. The weight of material removed was greater than any new building loads, so the risk of hill instability was not increased by the construction. The limestone from which the hill was formed showed signs of fissures and cavities, so remedial measures were necessary where these occurred below planned footings.

15. William Leonard Pereira (1909–1985) was an American architect from Chicago, Illinois, who helped define the look of mid-twentieth century America with his futuristic designs of landmark buildings including the Transamerica Pyramid in San Francisco, the Theme Building at Los Angeles International Airport (LAX) and the Los Angeles County Museum of Art (LACMA).

16. Philip Cortelyou Johnson (1906–2005) was an American architect best known for his modern architecture including the Glass House in New Canaan, Connecticut, and his postmodern architecture such as 550 Madison Avenue, New York, and the U.S. Bank Building (190 South LaSalle Street) in Chicago.

17. I.M. Pei (1917–2019) was a Chinese-American architect whose designs included the John F. Kennedy Presidential Library in Massachusetts, National Gallery of Art in Washington, D.C., Bank of China Tower in Hong Kong, Museum of Modern Art in Qatar and the glass and metal pyramid for the Louvre in Paris.

18. San Francisco (1985)–there are now 11 Arup offices operating in U.S.

19. Negarestan Cultural Center, Tehran, was to be located at the Marble Palace, one-time residence of Reza Shah Pahlavi. A theater, studios, workshops, a library, bookstore, exhibition halls and offices would be located beneath an existing, established garden of pine trees that would maintain the urban area's green space. The octagonal pattern of the skylit foyer roof at ground level would become a major landscape element on axis with both the Pahlavi and the Negarestan Museum. The form of the underground theater would be replicated by that of the open-air amphitheater above. Below-ground facilities included a multi-level meeting hall, delineated by wedge-shaped light scoops at garden level. Because the traditional Iranian teahouse was once a cultural center, this area could also function as a restaurant—with architecture and choice of building materials to suit. What we had with Negarestan Cultural Center was a strong geometric pattern of brick, tile and timber used for pools, vaults, domes and skylights. These elements conferred to the building's contemporary concrete technology a unifying and visually appealing character of traditional Iranian architecture.

20. Richard McGarrah Helms (1913–2002) served as director of central intelligence (DCI) under President Lyndon Baines Johnson and President Richard Milhous Nixon. On his re-election in 1972, Nixon informed Helms that his services would not be required in the new administration. Nixon offered Helms the ambassadorship to the Soviet Union. Helms proposed Iran instead. He served as ambassador to Iran from 1973–1977 in what would be his last government post.

21. Ronald Louis Ziegler (1939–2003) was White House press secretary and assistant to the President during President Richard Nixon's administration.

22. Richard Baskcomb Haryott (born 1941) had, prior to 1976, been the Ove Arup & Partners director responsible for design and supervision of structural, civil and services engineering for the $100 million National Exhibition Centre (NEC) at Birmingham, UK.

Chapter 13

1. Sir Edmund ("Ted") Happold (1930–1996) was a structural engineer and founder of consulting engineers Buro Happold. As a graduate of Leeds University, he joined Ove Arup & Partners on the recommendation of architect Basil Spence (at Arup he worked initially with Povl Ahm, engineer for St. Michael's Cathedral in Coventry). In 1959 Happold moved to New York to work with civil engineer Fred Severud (known for Madison Square Garden and the Gateway Arch in St. Louis). Happold returned to London to work with Arup in 1961. He worked on landmark buildings including Sydney Opera House and Centre Pompidou. He collaborated with Frei Otto, setting up a laboratory to study lightweight tensile structures with Ian Liddell, Vera Straka, Peter Rice and Michael Dickson. Leaving Arup in 1976 after Arup refused to allow him to start an office in Bath, he became professor of architecture and engineering design at the University of Bath and founded Buro Happold with seven Arup colleagues.

2. Initially, in 1976, Buro Happold offered only structural engineering consultancy, with a particular strength in lightweight structures. But in 1977 it added civil engineering and geotechnical engineering and, in 1978, building services engineering. By 1993 Buro Happold had 130 employees and eight partners, and in 1998, two years after Ted Happold died, the practice had grown to 300 employees and 12 partners.

3. Dr. Kamal El Kafrawi was, In January 1973, an Egyptian architect practicing in Paris. He was commissioned by UNESCO to carry out the preliminary study for the proposed colleges. A policy report subsequently prepared (May 1974) by the college of education in Qatar defined the need for a facility to accommodate 4,650 students and several disciplines. The idea was that the proposed colleges of education would make up a faculty of education that would be co-located with faculties of science, humanities & social sciences and Sharia & Islamic studies. Dr Kafrawi suddenly had a university project on his hands.

His envisaged facilities would have to be scaled up. But the 1973 design philosophy would be applied to the expanded proposal. This philosophy derived from Dr. Kafrawi's talks with a wide range of people in Qatar, his considerations of the traditional ways of life in the state and his studies of the construction of the older buildings of Doha, Wakrah, Khor and the North City.

What emerged was the view that holding fast to the traditional ways of building in the Islamic world, without change or improvement, would be a mistake—it would prevent the Arab from enjoying the benefits of modern technology. Dr. Kafrawi considered, however, that it would also be a mistake to import directly the building forms of Europe and America into the Islamic state of Qatar—these could be disorientating and disturbing to Islamic values. What he wanted to do was use Western technology to develop the existing building forms of the Arab culture. He wanted to generate an ongoing and continuously improving process of facilities design and construction which confirmed always strong roots in Arab civilization.

Dr. Kafrawi knew this project would be significant not only in Arabian Gulf terms but also within the global context. Unusually for the time, he didn't just want to arrive at a fixed solution. He wanted his work to stimulate discussion about the old and the new, present and future, national and international.

4. Brown, Mike and Chris Barber, "University of Qatar: Phase 1a," *The Arup Journal,* Vol. 21, No. 4, Ove Arup Partnership, Winter 1986, 2–9.

5. The concept for high quality university buildings in a modular low-rise form allowed the use of repetitive precast elements for both cladding and structural walls. To achieve the thermal response characteristics required, we chose to use concrete as the structural material throughout. We established a precast factory close by the site to achieve fast production of high-quality concrete units. This plant turned out 1 million blocks, 1 million paving slabs and tiles and 25,000 other elements of construction for Phase 1 of the project. These products were of exceptional quality—comparable with the finest produced anywhere in the world.

6. Sir Michael Terence Wogan (1938–2016), better known as Terry Wogan, was an Irish radio and television broadcaster who worked for the BBC in the UK for most of his career. Wogan was a leading media personality in Britain and Ireland from the late 1960s and was often referred to as a "national treasure."

7. Prior to design work, Dr. Kafrawi had studied the local factors which dictate the forms of buildings to codify them for the design. His consideration of natural ventilation focused on the 'Towers of Wind,' the traditional form of natural ventilation found throughout the Gulf region. These became the basis of the ventilation in the university lecture rooms and residences, although we adopted a modern method of control. The Towers of Wind are not only a substitute for mechanical ventilation and air-conditioning, in the event of power failure, but are also the defining shape which characterizes the outline of the university buildings and relates them to their cultural environment. Enhanced air circulation from the Towers of Wind has the additional benefit of modifying the effects of high humidity.

In these ways the architectural design facilitated natural ventilation and shading from the heat of the sun, creating acceptable living and working conditions over a proportion of the academic year. During the hottest part of the year the natural ventilation openings are closed, and mechanical air-conditioning takes over. We designed a ducted all-air system for larger areas, such as laboratories and lecture theaters, and used fan coil units for classrooms, studies and staff rooms. Each system is provided with dust filtration of the recirculated air and incoming fresh air. To avoid unnecessary wastage of water, we designed the central cooling plant to operate on the principle of air cooling.

Deriving the building form and ventilation from local tradition meant that we could now draw on hundreds of years of associated Arab natural lighting design development. In the UK I'd been concerned with getting daylight into buildings, but the concern in the Gulf was with controlling the strength and direction of the brilliant, penetrating sunshine. Dr. Kafrawi used specially oriented openings and screens in carved timber (mashrabiyyas) or gypsum to fill window openings and reduce light intensity. He designed surrounding circulation areas and stairways as shading devices and used diffuse overhead lighting from openings in the roofs of large spaces.

The design of the window openings prevented most direct solar radiation from entering the buildings. By allying window design with substantial wall thickness, we could simply and economically gain additional control of natural temperature (traditional Islamic buildings have walls ranging from 0.6 m [2 ft] to 1.5 m [5 ft] thick in order to achieve adequate insulation).

As a result, we were able to implement Dr. Kafrawi's concept of using traditional Arab shapes for the building superstructures. Lecture rooms are octagonal in plan, 8.4 m (27.5 ft) across and linked to at least two lobbies 3.5 m (11.5 ft) across. The first lobby serves as an entrance, small library for the lecture room and transition space between exterior and interior (to smooth the change from air-conditioned room to outside climate). The second lobby is a source of natural light and a space for use by small groups of students in discussion or study.

The octagonal lecture room has a roof structure created from sloping, precast units arranged to transpose the octagon plan module into a square. This forms the base of the Tower of Wind structure, which is characterized by its vertical slatted precast panel sides. The combination of polygonal form and multi-pitched roof reduces exposure to direct radiation of any one surface because the structure shades itself. This limits heat absorption by each surface, thereby limiting the transfer of heat to the interior. The octagonal plan, with domed roof and square lobbies, dampens echoes so it can provide good acoustics without the expense of sound insulation.

Variations of roof structure include the combination of two octagons to produce a laboratory building, in which case we used Vierendeel precast concrete roof trusses to span the double octagon. Where four octagons are combined as courtyards, we adopted a cast insitu folding-roof geometry.

8. The open-air ceremony was to take place in the early evening in the impressive piazza between the university library and the administration buildings. This could turn out to be a

mistake because the day began with a howling gale which posed many problems for the organizers and raised the probability of a sandstorm.

Beautifully printed invitations had been sent out with, in the corner, the traditional Arab octagonal shape that had been used so extensively in the university planning. The octagons on the Arup invitations were green, signifying that we were to be seated in the zone designated 'Important Senior Employees of the Nation.' Approximately 3,000 guests had been invited, so we were surprised, and very honored, to find ourselves sitting amongst the various attendee ambassadors, very close to the Amir's podium.

There was plenty to hold the attention as the wind played havoc with the sound systems. Teams of technicians tested not only the sound but also the patience of the high-level security enforcers. Eventually, the Amir and his entourage arrived and settled into immense leather armchairs on the podium. A series of speeches in Arabic ensued. These were greeted with enthusiastic applause. Mike Brown adopted his dourest Scottish persona to whisper to me, "The crowd's discovered a way to keep warm." Following the speeches, the president of the university presented the Amir with an octagonal gold medal, and a commemorative octahedron stone was unveiled. Then the Amir and his delegation set off on a tour of the university buildings, leaving guests to seek refuge in the library, where brightly coloured drinks and sticky buns were available.

The 17 Arup Doha site staff, reinforced by me and four other Arup attendees from London, didn't want the festivities to stop there—this day did, after all, also mark 10 years of Arup work in Qatar. We had a drink at Neil Noble's villa, watched a video replay of the ceremony on TV and headed for the Ghanem Garden complex with 180 or so guests. For months Arup Doha staff had been laying aside part of their monthly alcohol allocation in anticipation of such a party. Furthermore, an inspired act of project planning at start on site had ensured a case of champagne for broaching on this day.

Early in the evening we were honored by a visit from Mr. Hisham Qaddumi, Technical Director of the project for His Highness the Amir. Then we played host to Dr. Kafrawi and his wife Nadia. They were in very good form, having been congratulated by the Amir and having enjoyed excellent reactions from all the high-profile visitors to Doha for this day.

Our impromptu party continued until 2 a.m. The Wollopers darts team (Widnell and Trollope in consortium with Ove Arup & Partners) was the last to leave—some things in Doha never change. I enjoyed a final view of the university campus shortly after takeoff that day and spared a thought for the Arup Doha team which had reported to site at 7:30 a.m. for the next phase of the university.

Chapter 14

1. Amir-Abbas Hoveyda a.k.a., Hoveida (1919–1979) was an Iranian economist and politician who served as Prime Minister of Iran from January 1965 to August 1977. He was prime minister for 13 years and is the longest-serving prime minister in Iran's history. Because of the departure of the Shah and much of the ruling class, Hoveyda became, in 1978, the most prized prisoner of the Iranian Revolution. He was tried by the newly established Revolutionary Court for "waging war against God" and "spreading corruption on earth" and was executed.

2. Gholam-Reza Nikpey a.k.a. Nikpay (1927–1979) was deputy prime minister of Iran and mayor of Tehran. He became mayor of Tehran in 1969 and, prior to that, he had served as Iran's Minister of Housing from 1966 to 1969. During his tenure as the housing minister, an earthquake rocked the Province of Khorasan, causing mass destruction. He was in charge of rebuilding. It turned out to be one of the best rebuilding projects in the country's history. In 1977, he was appointed to Iranian senate by the Shah. He was executed on April 11, 1979, on the orders of a revolutionary tribunal, without legal representation or a chance to defend himself.

3. "Staddle stones (a.k.a. steddle stones) were originally used as supporting bases for facilities such as granaries, hayricks and game larders. The staddle stones lifted the granaries above the ground, thereby protecting the stored grain from vermin and water seepage. In Middle English "staddle" or "stadle" is "stathel", from Old English "stathol", a foundation, support or trunk of a tree. They can be mainly found in England, Norway ("stabbur"), Galicia and Asturias (northern Spain)." (Wikipedia).

4. Ferdinand Emmanuel Edralin Marcos, Sr. (1917–1989) was a Filipino politician who became the tenth President of the Philippines from 1965 to 1986. A leading member of the New Society Movement, he ruled as a dictator under martial law from 1972 until 1981.

5. Imelda Romualdez Marcos (born 1929) is a Filipino politician who was First Lady of the Philippines for 21 years. She married Ferdinand Marcos in 1954 and became First Lady in 1965 when he became president of the Philippines. During her term, she initiated numerous grand architectural projects using public funds.

6. In 1782 plans were made for a chamber of commerce in the Cornhill part of the City of London. Proposals included an information office for business and trade enterprises to use for "consultation, opinion, advice, information and assistance." A subscription rate of not less than three guineas, a significant sum at that time, was suggested for the service. But it took another 99 years of discussion before London had a chamber of commerce of its own. Finally on July 25, 1881, the London Chamber of Commerce and Industry

was established at Mansion House in the City of London with 130 members. Some of these firms or their successors still play a role in London Chamber today. From its earliest days the London Chamber's prime role has been in developing international trade and representing the interests of the London trading community, including assisting members in resolving day-to-day trading issues.

7. The Kylesku car ferry service became a bottleneck on the route to central Scotland and was replaced by the Kylesku Bridge in 1984. Opened by H.M. The Queen, the bridge is 276 m (906 ft) long and crosses a 130 m (430 ft) stretch of water. The curving, five-span, continuous, pre-stressed concrete, hollow bridge is one of the most beautiful bridges in the world. The Highland Regional Council had initially asked Ove Arup & Partners Scotland to prepare a feasibility study for a bridge, and this was ultimately designed by an Arup team led by Jorgen Nissen.

8. A rondavel is a traditional African dwelling with a conical thatched roof. It is usually round or oval in shape and made with materials that can be found locally in raw form. Its walls are often constructed from stones. The mortar may consist of sand, soil, or combinations of these, mixed with cow dung.

Chapter 15

1. Wigmore Street is dominated by Victorian architecture, but it is an Edwardian baroque building, the Debenham and Freebody Department Store, which stands out. Its dazzlingly white appearance is down to its Doulton's Carrara tile cladding. With more than a century of service, the tiles have lived up to the manufacturer's guarantee of smog-defying brilliance. The building is detailed with bare-breasted statues denoting virtues of commerce and industry and is further adorned with wreaths, garlands and angels, surmounted by a full attainment of the Royal Arms given pride of place on the central tower.

Chapter 16

1. A theater was first proposed when South Hill Park became an arts center in 1973. It didn't become a reality until the construction of the Wilde Theatre 1982–4. This courtyard theater, as used in Shakespeare's time, was a small, low-cost project, funded by a partnership of the local community and local businesses. There was, because of cost constraints, never any intention to build in the existing style of the historic South Hill Park, which dates back to the seventeenth century. The theater was, however, clad in red bricks to harmonize with those used in the main house.

2. Now known as Nottingham Royal Concert Hall, work began in 1980 and was completed in 1982, providing Nottingham with a contemporary 2,499-seat auditorium. The first artist to perform there was Elton John in November 1982.

3. Ed Mirvish (known as "Honest Ed" Mirvish), bought the theater for £550,000 (U.S.$962,000 at 1982 exchange rates), sight unseen, in August 1982. Ed was convinced that the theater's reputation would make it a viable commercial venture. Specifically, he wanted to restore it to the appearance of the J. T. Robinson design of 1869. One of my challenges was that the building owner, Ed, wanted to minimize cost, but my client, RHWL, wanted to achieve an exceptional result. We managed to avoid potential conflicts by breaking down the project into its constituent parts such as auditorium, seating, curtains, stage foyers, dressing rooms, ancillary accommodation and external works. As accurate a budget as possible was attached to each constituent part. This enabled the relative value of each aspect of the works to be considered by the client and the design team, so that expenditure would be put to maximum effect throughout the project duration. This exercise led to identification of a 'minimum scheme,' costed at £1 million (U.S.$1.75 million at 1982 exchange rates), and a 'full response scheme,' costed at £2 million. Perhaps inevitably, the agreed solution represented a cost midway between these two. The intermediate scheme initially excluded items from the preferred scheme but gave the option to 'put back in' specific items by defined dates during the contract.

4. The maximum Noise Rating (NR) Level for concert halls, broadcasting and recording studios, and churches, is 25. Noise Rating (NR) is most commonly used in Europe while Noise Criterion (NC) is a similar system more common in U.S.

Chapter 17

1. Royal Academicians (RAs) are practicing artists who help steer the Royal Academy's vision, support its activities and plan for the future. Each Academician is elected by his or her peers in one of four categories: painter, sculptor, architect or printmaker. At any one time there are no more than 80 RAs. They are all practising, professional artists who work in the UK. RAs in the Architect category range from George Dance The Younger (1768) through to Norman Foster (1991), James Stirling (1991) and Zaha Hadid (2005).

Chapter 18

1. Patrick Edmund Pery, 6th Earl of Limerick (1930–2003) was an Irish peer, banker and public servant. He became a director of the merchant bank Kleinwort Benson, having helped during

the merger of Kleinwort with Benson. In 1970 Limerick became Under-Secretary of State for Trade in the Conservative Government 1970–1974. On the defeat of the Heath government in 1974, he became president of the Association of British Chambers of Commerce (now the British Chambers of Commerce, the BCC) in succession to Sir Robin Brooke. In that capacity Pat Limerick led the transformation of the ABCC and raised it to a level of national influence at which it was effectively competing with, and sometimes eclipsing, the much younger Confederation of British Industry. After his three-year term of office with the Chambers of Commerce was completed, Lord Limerick then became chairman of the British Overseas Trade Board 1979–1983, as well as being chairman of the British Invisible Exports Council from 1975 to 1991. He was chairman of the European–Atlantic Group from 1999 to 2001.

2. Prince Edward, Duke of Kent (born 1935) has carried out royal engagements on behalf of his first cousin, Queen Elizabeth II, for more than 50 years. He served as the United Kingdom's Special Representative for International Trade and Investment, retiring from that position in 2001. He is president of the All England Lawn Tennis and Croquet Club, presenting the trophies to the Wimbledon champion and runner-up.

3. The Better Made in Britain campaign for British manufacturing industry included calling upon Her Majesty's Government to introduce a mandatory labelling scheme with an easily identifiable logo for all British-made goods.

4. The miners' strike of 1984–85 was a major industrial action to shut down the British coal industry in an attempt to prevent colliery closures.

5. Sir Ian MacGregor (1912–1998) was British-born but gained a reputation in North America for his ruthless, no-nonsense approach to reducing costs in ailing businesses. He was responsible for diminishing the power of the British unions during the 1980s while presiding as chairman of both British Steel and the National Coal Board.

6. Robert Haslam, Baron Haslam (1923–2002), was a British industrialist and chairman of British Steel Corporation and British Coal. He had been recruited by Manchester Collieries as a trainee in 1944, starting as a pit boy and working underground for three years, including 18 months at the coalface. He was created a Life Peer in 1990.

7. Michael Ray Dibdin Heseltine, Baron Heseltine (born 1933) was Deputy Prime Minister of the United Kingdom and First Secretary of State (1995–1997), Secretary of State for Trade and Industry and President of the Board of Trade (1992–1995), Secretary of State for the Environment (1990–1992) and Secretary of State for Defence (1983–1986). Following his retirement from the House of Commons at the 2001 UK general election, he was awarded a life peerage, taking the title Baron Heseltine of Thenford in the County of Northamptonshire.

8. Kenneth Wilfred Baker, Baron Baker of Dorking (born 1934) was Home Secretary (1990–1992), Chancellor of the Duchy of Lancaster and Chairman of the Conservative Party (1989–1990) and Secretary of State for Education and Science (1986–1989). Lord Baker, like me, is an Old Pauline.

9. Edward Arthur Alexander Shackleton, Baron Shackleton (1911–1994) was a geographer, Royal Air Force officer and Labour Party politician.

10. Sir Ernest Henry Shackleton (1874–1922) was an Irish Antarctic explorer who led three British expeditions to the Antarctic. He was one of the principal figures of the period known as the Heroic Age of Antarctic Exploration. After Roald Amundsen won the race to the South Pole in 1911, Shackleton set out to cross Antarctica from sea to sea, via the pole, in what became the Imperial Trans-Antarctic Expedition of 1914–1917. Its ship, *Endurance*, became trapped in pack ice and was crushed before the shore parties could be landed. The crew escaped by camping on the sea ice until it disintegrated, then launching the lifeboats to reach Elephant Island and, ultimately, South Georgia Island, a stormy ocean voyage of 830 miles (1,330 km). This was Shackleton's most famous exploit. In 1921, he returned to the Antarctic with the Shackleton–Rowett Expedition but died of a heart attack while his ship was moored in South Georgia. At his wife's request, he was buried there.

11. "*Rules* was established by Thomas Rule in 1798, making it the oldest restaurant in London. It serves traditional British food, specializing in classic game cookery, oysters, pies and puddings." (Rules Restaurant).

Chapter 19

1. "Thomas Hopper (1776–1856) was an English architect much favored by King George IV and particularly notable for his work on country houses across southern England" (Wikipedia).

2. "The London Underground system's first tunnels were built just below the ground, using the cut and cover method. Later, smaller, roughly circular tunnels—which gave rise to the nickname, the Tube—were dug through at a deeper level" (Wikipedia).

3. "Sir William Emerson (1843–1924) was a British architect who was President of the Royal Institute of British Architects (RIBA) from 1899 to 1902. Most of his later work was in India, where his most familiar design was the marble clad Victoria Memorial Hall in Calcutta (1905 onwards) which has been described as 'Britain's answer to the Taj Mahal'" (Wikipedia).

4. "Sir Thomas Edwin Cooper (1874–1942) was essentially a Classical architect whose designs included Marylebone Town Hall and Library, the headquarters of the Port of London Authority (at 10 Trinity Square) and the offices of Lloyd's in Leadenhall Street, City of London" (Wikipedia).

5. Acrow was a sad loss to British industry because of its manufacture of equipment and plant—especially cranes—for building, agricultural and forestry trades since the 1930s. It designed and installed steel Bailey bridges (Acrow panel bridges) throughout the world and was a renowned supplier of pit props, scaffolding and formwork. During World War II Acrow's production was largely diverted to the war effort, making tank and aircraft parts, as well as formwork for the Mulberry Harbour, used in the D-Day landings.

6. Peter Stone was trained at The Architectural Association in Bedford Square at the same time as Richard Rogers. I first met Peter when I was the structural engineer working on Northwick Park Hospital, where he was the architect responsible for the design of the operating theaters. After Northwick Park he was the architect partner at LDW for York District Hospital, where he again asked for me to be his engineer. Peter became a good friend, and after York we worked together on both the major development at Hammersmith Hospital and then the large building at St Mary's, London. Peter was also the partner representing LDW on the hospitals for the shah of Iran. Previously, Peter had carried out some work in Athens, which is why he was selected by the Onassis Foundation to be their Architect for the Memorial Cardiac Surgery Center.

7. Robert (Bob) Trew was at that time director of the Overseas Hospitals Practice of LDW and was responsible for design and management of all LDW health facility projects outside the UK. Bob and I had been working together for a long time because he'd been the LDW associate in charge of the Northwick Park Hospital and Clinical Research Complex (Northwick Park was, in its time, the largest civil building ever commissioned in England, so it was the biggest project in the early part of my career). After Northwick Park, Bob was responsible for LDW health projects in Iceland, the Netherlands, Belgium, Australia, Cyprus, Singapore and many other countries.

Chapter 20

1. "The Lipizzan or Lipizzaner is a horse breed dating back to the sixteenth century and named after the Lipizza Stud of the Habsburg monarchy. The breed is closely associated with the Spanish Riding School of Vienna, Austria, where the horses demonstrate the haute école or "high school" movements of classical dressage, including the highly controlled, stylized jumps and other movements known as the "airs above the ground." The breed has been endangered many times by wars across Europe, including World War I and World War II. The rescue of the Lipizzans during World War II by American troops was made famous by the Disney movie *Miracle of the White Stallions*" (Wikipedia).

2. The 100 m (328 ft)-long central mall atrium has an 18 m (59 ft)-span roof structure which supports an area of the park. Each end and the entire northern side of the atrium roof are glazed, making it desirable that we avoided any substantial roof support structure passing through the glazed area.

To minimize the size of structure in the glazed area, we adopted a propped cantilever scheme. We designed arched concrete beams to extend across the atrium as far as the glazing, where the beam is propped by a pinned steel circular hollow section (CHS). The cantilever beams had 15 m (49 ft) back spans linked into the main structure to maximize the cantilever stiffness and allow the central section of the roof beam to be made as shallow as possible. At their roots the 500 mm (19.7 in)-wide cantilever beams are 2,100 mm (82.7 in) deep, reducing to 800 mm (31.5 in) at the center of the atrium. The geometry of the prop joint was offset from the thrust line, inducing some reverse bending into the beam.

The connections at each end of the prop, although architecturally detailed as pins, are simple bearing connections. Because the prop had to be in place at the time of casting the roof beams, we had it fabricated with bearing connections located at each end and held in position while the bottom connection was formed, and then the roof beam cast around the upper connection. The lower end was connected to a precast node cast into the main concrete structure. The upper end, a fabricated bearing plate incorporating cheek plates to assist in confining the concrete, was cast into the beam at the time of pouring the concrete. Because the concrete is fair-faced and highly visible, a high degree of accuracy was required in positioning the prop and casting the beam.

Castle Mall is another of my major projects that demonstrates the value of having a world-class, in-house fire safety design team. In this case the steel props are not formally fire protected. They could have been. They could have been filled with concrete, in which case a slightly bigger section would have had to be specified. Instead, our fire safety designers determined that the location of the props relative to the nearest potential fire source would allow them to be unprotected. The props are based 5 m (16.4 ft) above the closest concourse, and the shop fronts are fire protected from 2.1 m (6.9 ft) above floor level. Calculation of the flame temperature at such a distance from the source of the fire demonstrated that insufficient heat could be transmitted to the exposed steel to raise its temperature to a level that would significantly impair its load capacity.

This may read like a small aspect of a huge project. So it was. But the project is a shopping center, with all the attendant implications for the safety of pedestrians, including mothers with small children. And the shopping center is built into the ground. Successful fire safety design based on a strategic approach, rather than on an encasing-in-fire-retardant-materials approach, makes for good engineering and good architecture. It helps eliminate the ugly and promote the elegant.

3. We established an eight-level car park for 760 vehicles on the eastern side of the site (one of the planning requirements was that the project would more than replace the 500 surface car parking spaces lost by development). The main car park is mechanically ventilated with up to 120 m³ (4,248 ft³) per second of fresh air, using a novel form of air movement. To avoid complication, we supplied air at one end of the car park and extracted it at the other end, 120 m (394 ft) away. To avoid pockets of air stagnation we used a supplementary system of air nozzles—fed by small diameter ducts—to swirl the air through the car park. In addition, by locating car entry at the lowest point and car exit at the highest point, we encouraged natural ventilation by the stack effect.

4. To achieve a sustainable park, which would include trees, we needed soil depths averaging 1.2 m (4 ft) increasing to 2 m (6.6 ft) in some areas. The soil also had to be naturally drained. So we designed our support structure as a complex arrangement of slabs that would allow natural drainage of the soil but would not allow water penetration into the slab. The slab configuration was also designed to minimize soil loads on the structure, minimize the need for applied finishes, accommodate level changes and accommodate thresholds between external and internal spaces. Generally, a 750 mm (29.5 in)-deep waffle slab was used, but in some areas shallower 400 mm (15.7 in) flat slabs were used to meet the design criteria.

Chapter 21

1. "Sir Edwin Landseer Lutyens (1869–1944) was thought by many to be the greatest British architect of the twentieth century or, indeed, of any century. He designed many English country houses, war memorials (including the Cenotaph in Whitehall) and public buildings and played an instrumental role in designing and building New Delhi, which would later serve as the seat of the Government of India (New Delhi is also known as "Lutyens' Delhi"). It is less well known that, with consulting engineer H. Fitzsimons, Lutyens designed the Runnymede Bridge at Staines in 1939. Because of World War II, construction was delayed by 20 years (Lutyens died in 1944, but his colleague George Stewart served as consulting architect, adopting the 1939 design). In the 1970s a new bridge was required to carry traffic on the eastern side of the existing bridge. It was required that the new bridge complement the existing Lutyens Bridge, and this was carried out beautifully by an Arup team including Ove Arup and Bill Smythe" (Wikipedia).

2. "Giacomo Vignola (1507–1573) was one of the great Italian architects of the sixteenth century. His two finest designs are considered to be the Villa Farnese at Caprarola and the Jesuits' Church of the Gesù in Rome. The three architects who spread the Italian Renaissance style throughout Western Europe were Vignola, Sebastiano Serlio and Andrea Palladio" (Wikipedia).

Chapter 22

1. "Sir Denys Louis Lasdun CH, CBE, RA (1914–2001) was an eminent English architect best known for the Royal National Theatre, on London's South Bank of the Thames, which is a Grade II* listed building and one of the most notable examples of Brutalist design in the UK. Lasdun was influenced not only by Le Corbusier and Ludwig Mies van der Rohe but also by classical English architects, including Nicholas Hawksmoor (circa 1661–1736) a leading figure of the English Baroque style of architecture" (Wikipedia). Arup loved working with Lasdun, and their early collaborations included a Grade II listed block of flats at 26 St James's Place (1960), to replace a building bombed in the war, and The Royal College of Physicians, Regent's Park (1964), one of the very few post-war buildings to be granted Grade I listed status.

2. IBM on London's South Bank was Sir Denys Lasdun's final major architectural design. My team designed its reinforced concrete structure with columns, flat slabs and beams, with precast concrete external components and non-structural internal blockwork walls. Our building services design included demonstration and conference suites, data processing libraries and catering facilities. The project started on site in 1980 and was largely completed in 1984. The building was planned to provide 300,000 ft² (28,000 m²) of space for 1,000 staff. Ironically, the granting of Grade II listed status on June 22, 2020, has created development considerations that can't this time be overcome by appointing Denys Lasdun.

Chapter 23

1. Richard St. John Vladimir Burton (1933–2017) was a British architect, of Russian and

Irish ancestry, who co-founded the architectural partnership of Ahrends, Burton & Koralek (ABK) in 1961 with his friends Peter Ahrends and Paul Koralek. He is best known for the design of low-rise housing, hospitals and low-energy systems. He was the ABK senior partner in charge of the firm's design of the British Embassy in Moscow, which was completed in 2000. ABK closed its practice in 2012.

2. "Hungerford Bridge, opened in 1864, crosses the River Thames in London, and lies between Waterloo Bridge and Westminster Bridge. Owned by Network Rail Infrastructure Ltd. (who use its official name of Charing Cross Bridge) it is a steel truss railway bridge flanked by two more recent, cable-stayed, pedestrian bridges—the Golden Jubilee Bridges—that share the railway bridge's foundation piers" (Wikipedia).

3. "Richard George Rogers, Baron Rogers of Riverside (1933–2021) is an Italian-British architect...best known for his work on the Pompidou Centre in Paris, the Lloyd's building and Millennium Dome, both in London, the Senedd in Cardiff and the European Court of Human Rights building in Strasbourg. He is a winner of the RIBA Gold Medal, the Thomas Jefferson Medal, the RIBA Stirling Prize, the Minerva Medal and Pritzker Prize. He is a Senior Partner at Rogers Stirk Harbour + Partners, previously known as The Richard Rogers Partnership" (Wikipedia).

4. "Sir Terry Farrell (born 1938) is a British architect and urban designer who graduated with a degree in architecture from Newcastle University, followed by a Masters in urban planning at the University of Pennsylvania in Philadelphia" (Wikipedia). He co-founded the practice Farrell/Grimshaw Partnership in 1965 and set up his own practice, Terry Farrell & Partners, in 1980. The firm's large-scale, new-build commissions in London included Embankment Place (1990), Alban Gate (1992) and Vauxhall Cross (1994). Farrell's Hong Kong office, incorporated as TFP Farrells, was founded in 1991. It was set up when the firm won an international competition to design the new Peak Tower, which opened in 1997 and was later featured on Hong Kong's $20 banknotes. TFP Farrells developed a strong reputation in urban transportation infrastructure beginning with the design for Kowloon Station (opened 1998) and the associated Union Square master plan, one of the largest air rights developments on Earth, which includes the tallest tower in Hong Kong, the International Commerce Centre (ICC). In mainland China, The KK100 tower in Shenzhen, completed 2011, is the tallest building to be designed by a British architect. The firm has won many awards for its mixed-use schemes, transit-oriented development, contextual urban placemaking and cultural buildings.

5. Norman Robert Foster, Baron Foster of Thames Bank (born 1935) was a member of the Team 4 architectural practice founded in 1963. He then founded Foster Associates (later Foster + Partners) which had major successes with the Willis Faber & Dumas Headquarters, Ipswich (1974); Sainsbury Centre for Visual Arts, University of East Anglia (1978); HSBC Main Building, Hong Kong (1985); Terminal Building at London Stansted Airport (1990); American Air Museum at Duxford (1997); Swiss Re Headquarters, London (2003); Millau Viaduct in southern France (2004) and Apple Park Corporate Headquarters, Cupertino, California, U.S. (2017). In 1999, Norman Foster was awarded the Pritzker Architecture Prize, often referred to as the Nobel Prize of architecture.

6. "Greycoat is an independent real estate company with expertise in the central London office market. It is a specialist in the development and asset management of large and complex projects" (Greycoat.com).

7. Marcus Binney, *Palace on the River: Terry Farrell's Redevelopment of Charing Cross*, Wordsearch Publishing, 1991, Introduction by Terry Farrell, Spring 1992.

8. Guy Battle (born 1962) and Chris McCarthy (born 1956) resigned from my Arup Building Engineering Group 5 in 1993 and founded Battle McCarthy Consulting Engineers in that year. They set out to provide a full range of engineering services, including structural engineering, environmental and building services engineering, electrical and mechanical engineering, public health engineering, civil engineering, energy systems design and environmental and landscape ecology design. Battle McCarthy was named Building Services Engineer of the Year 1995 and, in the same year, won the RICS award for Energy Efficient Building of the Year for the Ionica Building, Cambridge.

Chapter 24

1. Aintree is a Lancashire village best known as the site of Aintree Racecourse, which has since the nineteenth century staged the Grand National horserace. During the 1950s and 1960s, there was also a three-mile-long international Grand Prix motor racing circuit on the site, which used the same grandstands as the horserace. A shorter form of the racing circuit is still used for various motorsport events.

2. "The Great Storm of 1987 was a violent extratropical cyclone that occurred on the night of October 15–16, with hurricane-force winds causing casualties in the United Kingdom, France and the Channel Islands as a severe depression in the Bay of Biscay moved northeast" (Wikipedia).

3. The red telephone box, a public telephone kiosk designed by architect Sir Giles Gilbert Scott (1880–1960), has been a familiar sight on the streets of the United Kingdom, Malta, Bermuda and Gibraltar. The color red was chosen to make it easy to spot. "From 1926 onwards,

the fascias of the kiosks were emblazoned with a prominent crown, representing the British government. In 2006 the K2 version telephone box was voted one of Britain's top 10 design icons along with Concorde, the London Underground Map, Supermarine Spitfire aircraft, Mini car, World Wide Web, Routemaster Bus," Cat's eye (road), Tomb Raider album cover and Grand Theft Auto video games (Wikipedia).

Chapter 25

1. "The Muthaiga Country Club, near Nairobi, opened on New Year's Eve in 1913 and became a gathering place for the colonial British settlers in British East Africa, which later became, in 1920, the colony of Kenya...The Muthaiga Club is a setting in Beryl Markham's 1942 memoir *West With the Night*. It features in Ernest Hemingway's posthumously published (1970) novel *Islands in the Stream* and in Evelyn Waugh's 1931 travel book *Remote People*. Waugh [unlike me!] was unable to find accommodation on the premises" (Wikipedia).

Chapter 26

1. "The United States Army Corps of Engineers (USACE) is a U.S. federal agency under the Department of Defense and a major Army command of some 37,000 civilian and military personnel, making it one of the world's largest public engineering, design and construction management agencies" (Wikipedia).

2. "The Special Air Service (SAS) was created by Lieutenant Colonel David Stirling (1915–1990) in 1941 as a special forces unit of the British Army. It was conceived as a desert raiding force and operated initially behind German lines in north Africa, carrying out sabotage missions and wreaking havoc along Rommel's supply lines" (Elite UK Forces). Today the unit's work includes covert reconnaissance, counter-terrorism, direct action and hostage rescue. Because of the sensitivity of its operations, much about the SAS is highly classified and is not commented on by the British government or the Ministry of Defence.

3. A private automatic branch exchange (PABX) is an automatic telephone switching system within a private enterprise. Originally, such systems were called private branch exchanges (PBX) because they required a live operator, but almost all private branch exchanges today are automatic, hence PABX.

Chapter 28

1. Tim left me in 1995 to become managing director and vice president of business development at Tarmac, Black & Veitch in the UK. He returned to Canada in 2000, becoming co-founder and CEO of BIOX Corporation, a renewable energy company. He is now president and CEO at Mara Renewables Corporation, CEO at Valent Low-Carbon Technologies and CEO at FORGE Hydrocarbons Corporation.

2. It is a testament to Ron and to the Arup project directors liaising with him that, when he left Baghdad, Arup clients in Iraq were holding tender documents of some £600 million (some U.S.$910 million at 1983 exchange rates) value of construction work for which Arup was designer and which, it was anticipated at the time, would be built when the Iran–Iraq War ended. (Ron was also a fine rugby player before injury curtailed his sporting exploits.)

3. Anne Kriken Mann is currently a board member at the Institute of Classical Architecture & Classical America (ICA&CA), a U.S. non-profit organization dedicated to advancing the practice and appreciation of the classical tradition in architecture and the allied arts. On the UK side of the pond, she is a board member at the British Architectural Library Trust (BALT), a not-for-profit organization that was formalized in 2013 to support the cultural program of the Royal Institute of British Architects (RIBA).

4. "The Confederation of British Industries was formed in 1965 from a merger of the Federation of British Industries (FBI), the British Employers' Confederation and the National Association of British Manufacturers. Today it represents approximately 190,000 businesses made up of around 1,500 direct members and 188,500 non-members (non-members include some 140 trade associations)" (Wikipedia).

5. The British Consultants Bureau (BCB), founded in 1967, was the successor to the British Overseas Engineering Services Bureau that had been formed in 1966. BCB became the British Consultants and Construction Bureau (BCCB) in 2000, through the merger of the British Consultants Bureau and the International Construction Group. In 2006, BCCB changed its name to British Expertise. It promotes British companies and their professional services internationally.

6. Prince Richard, Duke of Gloucester (born 1944), the youngest grandchild of King George V and Queen Mary, practised as an architect before inheriting his father's dukedom in 1974. The Duke is Her Majesty The Queen's cousin and a full-time working member of the Royal Family. Many of The Duke's Patronages are related to his fields of special interest, including architecture and conservation.

7. Naruhito (born 1960) was titled His Imperial Highness The Prince Hiro (1960–1991) and then His Imperial Highness The Crown Prince of Japan (1991–2019) before succeeding his father Akihito as His Majesty The Emperor in 2019.

8. "The British Council for Offices (BCO) was established in 1990 and is Britain's leading forum for discussion and debate about the

issues affecting the office sector. Its mission is to research, develop and communicate best practice in all aspects of the sector. Its members are organizations involved in creating, acquiring or occupying office space including architects and construction industry professionals, lawyers, surveyors, financial institutions and public agencies. It advances the collective understanding of its members, enabling them to work together to create more effective office space" (bco.org.uk).

9. "The British Overseas Trade Board, replacing the British Export Board, was the export promotion agency of the UK Department of Trade and Industry from 1972 to 1988" (Wikipedia). It advised the British government on trade and export issues, conducted road shows to promote British exports and provided consulting services to new companies.

10. Dr. Andrew Chan (born 1949) graduated from the National Taiwan University in 1971 in civil engineering, received his doctorate from the University of Cambridge in 1975 in soil mechanics and spent the formative years of his career working in UK. He has led and been involved in innovative and award-winning projects such as the Hongkong Bank Building, IFC and ICC towers in Hong Kong, the 2008 Beijing Olympic venues, as well as major transport infrastructures including mass transit systems in eight cities, airports in Hong Kong and Beijing, roads and bridges, sustainable urban development, and energy projects throughout Asia. Dr. Chan was Deputy Chairman of Arup Group and, before then, Chairman of Arup's East Asia Region and leader of the Arup Hong Kong office. He is now Chairman of the Trustees' Board of the global Arup Group (Arup).

11. Sir Gordon Wu Ying-sheung (born 1935) is the chairman of the board of Hong Kong-listed Asian infrastructure firm Hopewell Holdings Ltd. which has interests spread across property investment and development, power, hotel and hospitality businesses. He studied engineering at the University of Manitoba in 1953, then transferred to Princeton University, where he graduated with a Bachelor of Science degree in engineering in 1958. Wu was the managing director of Hopewell from 1972 to 2002. He was responsible for Hopewell's infrastructure projects in mainland China and southeast Asia and has been involved in the design and construction of many buildings and development projects in Hong Kong and mainland China. He is also the chairman of Hopewell Highway Infrastructure Limited, subsidiary of Hopewell listed on August 2003, and an independent, non-executive director of i-Cable Communications Limited (Wikipedia).

12. "Jiang Zemin (born 1926) is a Chinese retired politician who served as president of the People's Republic of China from 1993 to 2003. Jiang officially introduced the term 'socialist market economy' ending a period of ideological uncertainty and economic stagnation following the Tiananmen Square protests of 1989. Under Jiang's leadership, China experienced substantial economic growth together with the continuation of reforms, the peaceful return of Hong Kong from the United Kingdom in 1997 and of Macau from Portugal in 1999, improving its relations with the outside world while the Communist Party maintained its tight control over the government" (Wikipedia).

13. The Queen's Award for Export Achievement is one of the highest official UK awards for British businesses. To be successful a company must show a substantial and sustained increase in export earnings over three consecutive 12-month periods, to a level which is outstanding for the products and services concerned and for the size of the organization. Awards are made on the advice of the Prime Minister after examination of applications by an advisory committee composed of leading individuals from industry, commerce, trade unions and government. The awards are conferred by Her Royal Highness Queen Elizabeth II on her birthday, which is on April 21.

Chapter 30

1. Swanke Hayden Connell Architects (SHCA) was founded in New York in 1906 as Walker & Gillette (the practice went through several name changes before becoming Swanke Hayden Connell Architects in 1981). The firm is best known for its buildings in Manhattan, New York City, which include Trump Tower, Continental Center and Americas Tower. Other notable projects include the Statue of Liberty Restoration on Liberty Island and the Eurasia Tower in Moscow. Swanke Hayden Connell Europe was acquired by British architectural practice Aukett Associates in 2013. SHCA filed for bankruptcy in 2015 over non-payment of fees issue by a client.

2. Richard Hughes is a trained engineer and building conservator who has undertaken projects for international and national organizations including the Aga Khan Trust for Culture, UNESCO/UNDP, ICOMOS, Ove Arup & Partners and the Egypt Exploration Society. His specialties are structural evaluation and conservation of historic buildings and archaeological sites. He gives scientific advice to firms of engineers and architects on engineering practices relating to historic sites, as well as correct use of traditional building materials, especially wood, soil and stone. He is internationally known for his work on traditional structures in hazard prone areas (affected by earthquakes and floods) and on the new use of soil as a structural building material. Over the past two decades he has conserved many historical wooden buildings in the Northern Areas of Pakistan and is a senior consultant

of the AKCS-P team that has won four UNESCO and British Airways conservation awards. He has been extensively involved with the science of in-situ preservation of archaeological remains (including sites in Britain and Mohenjo Daro) and has published widely on this subject (https://next.archnet.org/authorities/314).

3. In the 1990s "English Heritage was the operating name of an executive non-departmental public body of the British government, officially titled the Historic Buildings and Monuments Commission for England, that ran the national system of heritage protection and managed a range of historic properties" (english.heritage.org.uk).

4. "The Museum of London documents the social history of the UK's capital city from prehistoric to modern times. The museum is a few minutes' walk north of St Paul's Cathedral, overlooking the remains of the Roman city wall, on the edge the City of London. It is close to the Barbican Centre and is part of the Barbican complex of buildings created in the 1960s and 1970s to redevelop a bomb-damaged area of the City" (Wikipedia) (by architects Chamberlin, Powell and Bon working with Ove Arup & Partners).

5. "In the United Kingdom, a scheduled monument is a nationally important archaeological site or historic building, given protection against unauthorized change. Scheduling makes it illegal to undertake a great range of 'works' within a designated area, without first obtaining 'scheduled monument consent'" (Wikipedia).

Chapter 31

1. Richard Charles Albert Holbrooke (1941–2010) was an American diplomat who brokered the Dayton Accords (1995) to end the conflict in Bosnia and Herzegovina, served as U.S. ambassador to the United Nations (1999–2001) and was President Barack Obama's special representative to Afghanistan and Pakistan (2009–10).

2. "Francisco Javier Solana de Madariaga (born 1942) is a Spanish physicist and Socialist politician. He served in the Spanish government as Foreign Affairs Minister (1992–1995) as the Secretary General of NATO (1995–1999) and as the European Union's High Representative for Common Foreign and Security Policy, Secretary General of the Council of the European Union and Secretary-General of the Western European Union (1999–2009)" (Wikipedia).

3. "Wesley Kanne Clark, Sr. (born 1944) commanded Operation Allied Force in the Kosovo War during his term as the Supreme Allied Commander Europe (SACEUR) of NATO from 1997 to 2000 … Clark started the bombings codenamed Operation Allied Force on March 24, 1999, on orders to try to enforce U.N. Resolution 1199 following Yugoslavia's refusal of the Rambouillet Agreement" (Wikipedia).

4. "Bernard Kouchner (born 1939) is a French politician and physician who co-founded Médecins Sans Frontières (MSF) and Médecins du Monde. On July 15, 1999, pursuant to Security Council Resolution 1244, UN Secretary General Kofi Annan nominated him as the second UN Special Representative and Head of the United Nations Interim Administration Mission in Kosovo (UNMIK). During 18 months, he led UN efforts to create a new civil administration and political system, and to rebuild the economy shattered by the Kosovo War. He was awarded an honorary doctorate by the University of Pristina for his services to Kosovo" (Wikipedia).

5. Marc Franco (born 1947) is a Belgian who joined the European Commission in 1978, working initially on agricultural and food programmes, and macroeconomic structural adjustment programmes. From 1990–1998 he was responsible for sectors in the Phare programme, one of three pre-accession instruments financed by the European Union to assist the applicant countries of Central and Eastern Europe in their preparations for joining the EU. He later became Head of Unit responsible for relations with Hungary, Czech Republic, Slovakia and Slovenia. In 1998 Franco was appointed director of the Cohesion Fund and subsequently headed the EC Task Force for the reconstruction of Kosovo (1998–2000). In 2001 the EC appointed him Deputy Director General of Europe Aid–Cooperation. He is now a senior associate fellow at Egmont (The Royal Institute for International Relations), an independent think-tank based in Brussels.

6. "General Sir Michael David Jackson (born 1944) is one of Britain's most high-profile generals. He was appointed Commander of NATO's Allied Rapid Reaction Corps (ARRC) in 1997, leading them on deployment to Bosnia and Herzegovina in March 1999. Jackson commanded KFOR, NATO's multinational peacekeeping force established at the end of the Kosovo War. In 2003 he was appointed Chief of the General Staff (CGS), the professional head of the British Army. Jackson now speaks on military matters and works as a consultant and guest lecturer" (Wikipedia).

7. "Viktor Zavarzin (born 1948) became Russia's first military representative at NATO headquarters from November 1997 to November 2001. It was while in post at NATO following the Kosovo War that he led the 'dash to Pristina' that saw Russian troops, detached from the SFOR peacekeeping force in Bosnia-Herzegovina, arrive in Pristina before KFOR arrived there" (Wikipedia).

8. "General Francis Richard Dannatt, Baron Dannatt (born 1950) took command of 3rd Mechanised Division in 1999 and simultaneously commanded British forces in Kosovo. It was decided, given the large number of British troops serving as part of the multinational Kosovo Force (KFOR), that the 3rd Division's headquarters

would deploy to oversee British operations, with Dannatt serving as Commander of British Forces (COMBRITFOR). Dannatt was appointed Chief of the General Staff (CGS) in August 2006, succeeding General Sir Mike Jackson" (Wikipedia).

9. Jackson, General Sir Mike, *Soldier: The Autobiography of General Sir Mike Jackson*, Bantam Press, London ND, 2007.

10. "B&Q Limited is a British DIY and home improvement retailing company headquartered in Eastleigh, England, and founded by Richard Block and David Quayle in 1969 (originally as Block & Quayle)" (Wikipedia).

Chapter 32

1. "Vojislav Koštunica (born 1944) was the last president of FR Yugoslavia, from 2000 to 2003, and the prime minister of Serbia in two terms (from 2004 to 2007 and from 2007 to 2008). He won the 2000 Yugoslav presidential election as a candidate of a wide alliance Democratic Opposition of Serbia (DOS), which led to the overthrow of Slobodan Milošević and withdrawal of international sanctions against Yugoslavia" (Wikipedia).

2. "Zoran Đinđić (1952–2003) was the Mayor of Belgrade in 1997 and then the Prime Minister of Serbia from 2001 until his assassination in 2003. As Prime Minister, he advocated pro-democratic reforms and the integration of Serbia into European structures" (Wikipedia).

3. Karen Svensson, "Nigel Thompson's 'Outstanding Contribution,'" *The Bulletin*, News from around The Arup Partnerships, Issue 176, September 2001.

4. Sir Richard Francis Burton KCMG FRGS (1821–1890) was a British explorer, geographer, translator, writer, soldier, orientalist, cartographer, ethnographer, ethnologist, spy, linguist, poet, fencer, Freemason and diplomat. He traveled and explored in Asia, Africa and the Americas, experiencing many cultures and learning 29 European, Asian and African languages. "Burton's achievements include journeying to Mecca in disguise (at a time when Europeans were forbidden access on penalty of death), an unexpurgated translation of *One Thousand and One Nights* (commonly called *The Arabian Nights* in English), publication of the *Kama Sutra* in English, a translation of *The Perfumed Garden* (the Arab *Kama Sutra*) and a journey with John Hanning Speke as the first Europeans to visit the Great Lakes of Africa in search of the source of the Nile" (Wikipedia).

Chapter 33

1. "The Stability Pact for South Eastern Europe (1999–2008) aimed to strengthen peace, democracy, human rights and economy in the countries of South Eastern Europe. The Pact was made up principally of organizations outside the region and was replaced in February 2008 by a more regionally-owned co-operative framework, the Regional Cooperation Council (RCC)" (Wikipedia).

Chapter 35

1. Clough Williams-Ellis, *England and the Octopus*, Geoffrey Bles, London, 1928 (facsimile edition published by CPRE, 1996).

2. "The Campaign to Protect Rural England, now known as CPRE, the countryside charity, is a charity with more than 40,000 members and supporters campaigning for a sustainable future for the English countryside. CPRE is one of the world's longest-running environmental groups" (Wikipedia).

3. "Sir Max Hugh Macdonald Hastings, born 1945, is a British journalist and military historian, who has worked as a foreign correspondent for the BBC, editor-in-chief of *The Daily Telegraph* and editor of the *Evening Standard*" (Wikipedia). He will always be remembered for his incisive reporting when embedded with the British Forces during the Falklands War in 1982. Max has published 26 books.

4. "Department for Farming and Rural Affairs is now The Department for Environment, Food and Rural Affairs (Defra), the government department responsible for environmental protection, food production and standards, agriculture, fisheries and rural communities in the United Kingdom" (Wikipedia).

5. Consigia was renamed Royal Mail Group that same year, 2002.

6. "*Cold Comfort Farm—A Crisis for the Countryside* (1985):Central Independent Television documentary, producer / director / reporter Francis Gerard and reporter Max Hastings, looked at the crisis in British farming in which subsidy induced over-production and led to drastic falls in market prices, over-burdening of land and mass bankruptcies" (bfi.org.uk).

7. The Common Agricultural Policy (CAP), the agricultural policy of the European Union, is often criticized for its cost, environmental impact and humanitarian consequences.

8. A Green Paper is a government publication that details specific issues, and then points out possible courses of action in terms of policy and legislation.

9. Urban sites for potential building development that have hosted previous development.

10. Previously undeveloped sites.

11. A nimby (Not in my Back Yard) is a person who objects to the siting of something perceived as unpleasant or hazardous in the area where he or she lives, especially while raising no such objections to similar developments elsewhere.

12. "The Sustainable Communities Plan

was launched in 2003 and was a key policy of the Office of the Deputy Prime Minister in the Labour Government, guiding its regeneration and departmental objectives. It led to a range of policies and plans which were in effect a spatial plan for the whole of England" (Wikipedia).

13. "A greenbelt is a policy and land use zone designation used in land use planning to retain areas of largely undeveloped, wild, or agricultural land surrounding or neighboring urban areas" (Wikipedia).

14. "William McGuire Bryson, born 1951, is an American-British author of books on travel, the English language, science and other non-fiction topics. He wrote *Notes from a Small Island* (1995), an exploration of Britain, and its accompanying television series. His *A Short History of Nearly Everything* (2003), is widely acclaimed for its accessible communication of science" (Wikipedia).

15. Poundbury is designed to create a sustainable community which achieves an attractive, modern and pleasing place in which people can live, work, shop and play. Emphasis is placed on the quality of place-making through time-honored principles, urban design, landscaping and the selection of materials. Built on Duchy of Cornwall land, Poundbury is currently home to some 4,150 people in a mix of private and affordable housing and provides employment for 2,300 people working in more than 207 shops, cafés, offices and factories. An additional 550 people are employed in construction across the site, and many more are self-employed and at times work from home. (https://poundbury.co.uk/about/architecture-and-urban-design/).

16. Montagu Denis Wyatt ("Monty") Don, born 1955, is a British television presenter, writer and speaker on horticulture, best known for presenting the BBC television series *Gardeners' World*. In 2011 he presented *Monty Don's Italian Gardens*, a BBC2 series and, in 2013, a companion series, *Monty Don's French Gardens*. In 2018 he presented the BBC series *Monty Don's Paradise Gardens*, traveling across the Islamic world and beyond in search of paradise gardens and considering their place in the Quran. In 2019 he presented the BBC series *Monty Don's Japanese Gardens* and *Monty Don's American Gardens* series was broadcast in 2020 (Wikipedia).

Chapter 36

1. The Brandt Report is the report of the Independent Commission which was first chaired by former German Chancellor Willy Brandt. It reviewed international development issues and suggested primarily that, because such a great chasm in standard of living existed along the North-South divide, there should be a large transfer of resources from developed to developing countries (Wikipedia).

2. "Sir Seretse Goitsebeng Maphiri Khama, GCB, KBE (1921–1980) was the first President of Botswana (1966–1980). He was born into a royal family, of what was then the British Protectorate of Bechuanaland, and was educated in neighboring South Africa and in the United Kingdom. He married a British woman, Ruth Williams, which as a union between a black man and a white woman was controversial with the National Party government in South Africa, with the British and with the Bangwato. Khama founded the Botswana Democratic Party in 1962 and became Prime Minister in 1965. In 1966, Botswana gained independence and Khama was elected as its first president. During his presidency, the country underwent rapid economic and social progress" (Wikipedia).

Chapter 39

1. "The Morris Minor ('Moggie') was a hugely popular British car, with more than 1.6 million manufactured between 1948 and 1972 in three series: the MM (1948–1953), the Series II (1952–1956) and the 1000 series (1956–1979). Initially available as a 2-door saloon and tourer (convertible), the range was expanded to include a 4-door saloon in 1950, a wood-framed estate car (the Traveller) from October 1953 and panel van and pick-up truck variants from 1953. The Morris Minor was the first British car to sell over a million units, is a classic automotive design and has become a part of British culture" (Wikipedia).

2. "*The Vicar of Dibley* was a hugely popular British television sitcom which ran on the BBC,1994–1998, and then from 1999–2007. It was set in a fictional small Oxfordshire village called Dibley, which was assigned a female vicar (played by Dawn French) following the 1992 changes in the Church of England that permitted the ordination of women" (Wikipedia). It placed third in a BBC poll of Britain's best sitcoms.

3. "Lib-Dems is a colloquial term for the Liberal Democrats, a liberal political party in the United Kingdom" (Wikipedia).

Chapter 41

1. A thegn is a person ranking between an earl and an ordinary freeman, holding land of the king or a lord in return for services.

2. A rod is a unit of length equal to 11 cubits, 5.0292 m or 16.5 ft. A rod is the same length as a perch and a pole. The lengths of the perch (one rod) and chain (four rods) were standardized in 1607 by Edmund Gunter. The length is equal to the standardized length of the ox goad used by medieval English ploughmen; fields were measured in acres which were one chain (four rods) by one furlong (in the United Kingdom, ten

chains). As a unit of area, a square perch is equal to a square rod (30.25 square yards, 25.29 square metres or one one-hundred-and-sixtieth of an acre). The rod is still in use as a unit of measure in certain specialized fields. In recreational canoeing, maps measure portages (overland paths where canoes must be carried) in rods. This is thought to persist due to the rod approximating the length of a typical canoe. In the United Kingdom, the sizes of allotment gardens continue to be measured in rods. An allotment in the UK is a plot of land rented by an individual for growing vegetables or flowers (in U.S. an allotment is a piece of land made over by the government to a Native American).

Bibliography

Almonds-Windmill, Lorna, *A British Achilles: George, 2nd Earl Jellicoe KBE DSO MC FRS 20th Century Soldier, Politician, Statesman*, Pen & Sword, 2006.

Barrie, Malcolm, John Crack, Mark Facer, and Graham Phillips, "Embankment Place," *The Arup Journal*, Vol. 26, No. 1, Ove Arup Partnership, Spring 1991. https://www.arup.com/perspectives/publications/the-arup-journal.

Binney, Marcus, *Palace on the River: Terry Farrell's Redevelopment of Charing Cross*, Wordsearch Publishing, 1991.

Broomhead, Anthony D.W., and William J. Grose, "Discussion on The Structural Design of Castle Mall", *The Structural Engineer*, Vol. 73, Issue 6, The Institution of Structural Engineers, 1995.

Brown, Mike, and Chris Barber, "University of Qatar: Phase 1a," *The Arup Journal*, Vol. 21, No. 4, Ove Arup Partnership, Winter 1986. https://www.arup.com/perspectives/publications/the-arup-journal.

Campaign to Protect Rural England, *Annual Review 2005: Reporting on the Year 1 January–31 December 2004*, CPRE, June 2005.

Farrell, Terry, *Lives in Architecture*, RIBA Publishing, 2020.

Gates-Sumner, Rod Buchanan, "Lutyens House," *The Arup Journal*, Vol. 25, No. 2, Ove Arup Partnership, Summer 1990. https://www.arup.com/perspectives/publications/the-arup-journal.

Hastings, Max, "The Country Matters," *The Observer*, Sunday June 23, 2002.

Hirst, John, Roger Olsen, and Alan Pepper, "Merrill Lynch," *The Arup Journal*, Vol. 38, No. 1, Ove Arup Partnership, 1/2003. https://www.arup.com/perspectives/publications/the-arup-journal.

Home Office, *Air Raid Precautions: Directions for the Erection and Sinking of the Galvanised Corrugated Steel Shelter*, Crown Copyright, February 1939.

Al-Hussein, Queen Noor, *Leap of Faith: Memoirs of an Unexpected Life*, Weidenfeld & Nicolson, 2003.

Jackson, Mike, *Soldier: The Autobiography*, Bantam Press, 2007.

el Kafrawi, Dr. Kamal, *Preliminary Study for Proposed Colleges*, UNESCO, 1973.

King, Alice, "A Six-acre Garden of Vines," *House & Garden*, September 1985.

McCarthy, Christopher, "The Cast Steel Nodes for Lee House," *The Arup Journal*, Vol. 23, No. 1, Ove Arup Partnership, Spring 1988. https://www.arup.com/perspectives/publications/the-arup-journal.

Morreau, Patrick, "Royal Exchange Theatre," *The Arup Journal*, Vol. 11, No. 4, Ove Arup Partnership, December 1976. https://www.arup.com/perspectives/publications/the-arup-journal.

Pilkington, John, and Alan Foster, "The Old Vic Is Back!," *The Arup Journal*, Vol. 19, No. 1, Ove Arup Partnership, April 1984. https://www.arup.com/perspectives/publications/the-arup-journal.

Robertson, Jacquelin T., "Shahestan Pahlavi: Steps Toward a New Iranian Centre," *Toward an Architecture in the Spirit of Islam*. Renata Holod (ed.), Philadelphia: The Aga Khan Award for Architecture, 1978.

Thompson, Nigel, "Northwick Park Hospital," *The Arup Journal*, Vol. 4, No. 3, Ove Arup Partnership, September 1969. https://www.arup.com/perspectives/publications/the-arup-journal.

Wickham, Cynthia, "The Simple but Highly Decorative Life at the Mill," *House & Garden*, May 1969.

Williams-Ellis, Clough, *England and the Octopus*, Geoffrey Bles, London, 1928 (facsimile edition published by CPRE, 1996).

Copies of *The Arup Journal* listed above can be downloaded at: https://www.arup.com/perspectives/publications/the-arup-journal.

Index

Numbers in ***bold italics*** indicate pages with illustrations

Aalto, Alvar 56
Abu-Sitta, Salman Hussein ***27***, 32
Action for the River Kennet (ARK) 201–***202***, 203
Adams, Colin 154, 161
Afkhami, Sardar 57
Ahm, Povl 81–***82***
Air Rights Buildings 1, 105–***106***, 107–***108***, 109–114
Akhtamar, Lake Van, Turkey ***37***
Alban Gate, City of London 1, 3, 114–115
Allport, Peter 184
Anderson, David 61
Anderson, John 155
Anderson bomb shelter 3, 7–***9***
Arathoon, David 154, 164–165
Arup, Sir Ove 3, 5, ***30***–31, 50, 61, 81–***82***, ***84***
Arup Acoustics 1
Askew, Tim 155
Association of Civil Engineers' Outstanding Contribution Award 170, 225
Atkinson, Grahame 154
Atrium 2, ***98***–101
Aviation House, Gatwick, West Sussex 74
Aylard, Richard 201, 205
Azadi Freedom Tower, Tehran *see* Shahyad Monument, Tehran

Bailey, Mark 206.
Baker, Kenneth 85
Banda, Dr. Hastings 24, 38
Banks, Tony 197
Barber, Chris 96
Barker, Tom 61
Barrie, Malcolm 73, 95, 109–111, 186
Barron, Kevin 199
Baster, Jenny 81
Batchelor, John 134

Battle, Guy 115
Battle, John 154–***156***, ***159***
Baxter, Brian 91
Beckmann, Poul 3, 29–30, 33
Bennett, Diana ***11***
Better Made in Britain 85, 135–***136***
Blair, Tony 4, 154, 160, 173
Blake, Jeremy 185
Bolt, Vincent 199
Bonnett, Bridget ***102***, ***223***
Bonnett, Eustace (Jim) ***102***, 212
Bonnett, Philippa (later Webber) ***34***
Boys, Bruce ***126***–127
Britannic House *see* Lutyens House
British Airports Group (BAG) 137
British Consultants Bureau (BCB) 134–135, 137, 154
British Council for Offices 135
British Overseas Trade Board (BOTB) 85, 135
British Thai Business Group 137
Brown, Mike 61–***62***, ***63***–64, 91
Bryson, Bill 4, 193
Building Understanding Through International Links for Development (BUILD) ***195***–200
Business Development Unit (BDU) 131, 132–***136***, 137–***138***, 139–140
Byers, Stephen 154

Caborn, Richard 164–165, 166, 199
Campaign to Protect Rural England (CPRE) 2, 4, 188–***189***, 190–193
Castle Mall, Norwich 93–***95***, 96
Chan, Andrew 135
Clancy, Michael ***180***, 184

Clarke, Charles 199
Clayden, Ken ***82***–83
Coffin, Frank 28, 204
Cohen, Edwina (previously Bonnett) ***223***
Cohen family ***223***
Commander of the British Empire (CBE) ***138***–139, 225
computational fluid dynamics (CFD) 3, 92
Confederation of British Industries (CBI) 33; *see also* Federation of British Industries (FBI)
Construction Procurement Board 135
Cook, Don 155
Council for the Preservation of Rural England *see* Campaign to Protect Rural England (CPRE)
Council for the Protection of Rural England *see* Campaign to Protect Rural England (CPRE)
CPRE, The Countryside Charity *see* Campaign to Protect Rural England (CPRE)

Dannatt, Major General Richard 158–***159***
Dawson, Keith 81
De Vere, Roger 201–203
Diamond, Julian 132
Di Benedetto, Grace ***224***
Dixon, Joly 149, 155, 161
Doha University, Qatar 60–***62***, ***63***–65
Dowson, Sir Philip 81–***82***
Dunican, Peter 3, 5, 26, ***30***, 81, 84

Embankment Place, Charing Cross, London 1, 3, 105–***106***, 107–***108***, 109–114
Emmerson, Bob ***82***–83, 140

Index

Facer, Mark 73–74
Farrell, Sir Terry 1–2, 3, 75, 96, 105–*106*, 107–*108*, 109–*115*
Federation of British Industries (FBI) *24–27*; 225; see also Confederation of British Industries (CBI)
fire safety design 3
Foggo, Peter 39, *82*–83
Foster, Alan 75, 96
Franco, Marc 149–150, 161–163
Freeman of the City of London 225
French Cart, St. Lo *42*
Fuller, Major Joe 159–160

Garnica, Fernando 33, 35
Gates-Sumner, Martin 99
Gibson, Mark 154–155
Gill, Peter 61
Golden Square Shopping Centre, Warrington 50
Goodban, Dennis 17–19, 51–52
Gordon, Jeremy 12
Gordon, Mike 161
Graham, Sir Alexander Michael (Mike) 205
Grove House and Estate, Stitchcombe 49–52, 59, *214–215*, 216–*220*, 221–*222*, 223–224

Haddon, John 74, 145
Haig, Tim 80, 132
Hare, Nicholas 205
Harris, Andrew 187, *223–224*
Harris, Nim see Thompson, Nim
Harris, Tom 219, 221, *222–224*
Harrison, Luke *224*
Haryott, Richard 59, *82*–83
Hastings, Sir Max 4, 188–191, 193
Havergal, John 32, 34
Haviland Marshall & Bray 21–*22*, 23
Helms, Richard 58
Hemery, David 199
Henderson, Sandy 200
Heseltine, Michael 85, 135
Hill, Ann 32
Hill, Mike 32, *71*
Hirst, John 146
Hitchmough, Charlotte 201–*202*
Hobbs, Bob 28, 61, 81–*82*
Hobhouse, Penelope 219–*220*
Hollamby, David 183–184
Howard, Andy 144

Howard, Belinda 212
Hughes, Richard 146
Hurd, Douglas 64

Iacobescu, George 145
Institution of Civil Engineers 31; Fellowship 225; International Medal 166
Institution of Structural Engineers 31; Fellowship 225
Irons, Stuart 81–82

Jackson, Gen. Sir Mike 149, 155–158
Jellicoe, George, 2nd Earl Jellicoe 49, 53, 66, 69–70, 85, *86*
Johnson, Philip 56

Kafrawi, Dr. Kamal el 60–61, *63*–64
Kariba Dam 23
Kaunda, Kenneth 24, 38, 70
Kemp, Vic 28, *30*, *39*
Kennet, River see Action for the River Kennet (ARK)
Kinnock, Lord Neil 199
Knighthood 170–*171*–172, 225
Koch-Nielsen, Holger 56
Kosovo 2, 4, 151–*152*, *153*–156, *157–158*, *159–164*, 165
Kosovo Power Stations *158*–165
Kouchner, Dr. Bernard 155, 164
Kriken, Anne see Mann, Anne
Kuwait City 1991: Department of Trade and Industry (DTI) Task Force 124–*125*, *126*–128

Lasdun, Sir Denys 104
Law, Margaret 50
Lee House see Alban Gate
Leopard Rock Hotel, Rhodesia 21–*22*
Lester, Jim 197–198
Levinson, Keith 154, 160
Levitt Bernstein 50–51, 75
Lewis, Michael (Mick) 55, 81–*82*, 84
Livery of the Worshipful Company of Engineers 225
Llewelyn-Davies Weeks (LDW) 31, 38–*39*, 50, 53, 65, 88–92
London Aquarium 137
London Blitz 3
London Chamber of Commerce and Industry (LCCI) 69–70
Luke, Steven 186

Lutyens House, Finsbury Circus, City of London 97–*98*, 99–101
Lyall, Iain 145

MacGregor, Sir Ian 85
Major, Royal Engineers' Logistic Staff Corps. 166, 225
Manchester Royal Exchange Theatre 50–*51*, 75
Mann, Anne 79–80, 132, 145
Manning, Martin 56
Marcos, Pres. Ferdinand 69, *86*
Marlborough Brandt Group, The *194–195*
Marsh, Ron 132
Martin, John 81–*82*, 107
Maurice, Dr Nick *194–195*, 196–198
McCarthy, Chris 115
Mercers Company 204
Merrill Lynch, Newgate, City of London 146–148
MI6 Headquarters Building, Vauxhall Cross, London *115*
Michael, Sir Duncan 54, 81–*82*, 139
Milburn, Alan 160
Milburn, Roger 91
Milosovic, Slobodan 4, 151, 166, 176
Minal Parish Council 210–211
Montenegro see Serbia and Montenegro
Montgomery, Field Marshall Bernard Law 13
Morrison bomb shelter 3, 7
Moynihan, Lord Colin 70

National Hospital for Nervous Diseases, London 49
National Recreation Centre, Crystal Palace, London 3, 28–*29*
Needham, Richard 135
Night Blight CPRE Campaign 192
Nikpay, Mayor of Tehran *55*, 58, 66
Nissen, Jorgen *82*–84
Nkomo, Joshua 24
Noble, Neil 61
Noor, Queen Noor of Jordan 55–56
Northwick Park Hospital, Harrow, London *31*, 38–*39*
Nottingham Concert Hall 69

Oberoi Hotel Group 181
Old Vic, The 69, 75–*76*, 77
Olley, Julian 96

Onassis Cardiac Surgery Center, Athens 89–91
O'Sullivan, Paul 185

Pei, I.M. 56
Pereira, William 56
Perry, Brian 52
Peter Inskip + Peter Jenkins *98*–101
Pilkington, John 75, 91, 96, 115
Pope John Paul II 151
Price, Sam 33, 41
Prince of Wales, HRH Charles 100, 170–*171*, 193, 201
Prince Richard, Duke of Gloucester 134
Prior, Lord Jim 135

Queen Elizabeth II 70, *138*–139, *189*, 193
The Queen's Award for Export Achievement 84, 137–*138*

Rachman, Adam 209
Renton Howard Wood Levin (RHWL) 60, 75–76
Reporter, Sir Shapur 54
Restorative Development 4
Rhodesia 18–*19*, *20*–*22*, *23*–*24*, 25
Rice, Peter 50, *82*–83, 137
Rigby, Roger 61, 81–*82*
Rivera, Alberto Jesus 33
Robertson, Jaquelin T. *55*, 58
Robertson, Johnny 206
Rogers, Lord Richard 30, 107
Rowlands, Sir David 205
Russell, Bruce 155

St. Helena 2, 19, 178–*179*, *180*–187
St. Helena Leisure Corporation (SHELCO) *179*–*180*, 181–186
St. Mary's Hospital, Paddington, London 88–*90*
St. Paul's School 13–*14*, *15*–16, 204–209
Salisbury Polytechnic, Rhodesia 21, 25
Saunter, Will 207
Schneider, Dux 32, 36–37, 45
Schneider, Monique 32, 36–37, 45
Serbia and Montenegro 166–*169*, 170
Shah of Iran, Mohammad Reza Pahlavi 53, *55*
Shahestan Pahlavi, Tehran 55–*56*
Shahyad Monument, Tehran *54*

Sharjah General Hospital, United Arab Emirates 92
Shears, Mike *82*–83
Shemie, Sam 146
Short, Clare 160, 162
Singapore British Business Council 137
Sobieski, Therese 155
Southwood, Bill 146
Stability Pact for South-Eastern Europe 173–177
Stephen, Dr. Martin 205
Stitchcombe Vineyard 67–*68*, 86–87, 102, 116
Stone, Brian 161–163
Stone, Peter 39
Stonehenge 193
Street, Peter 161
Summers, Royston 39
Sutcliffe, Robin 31
Sydney Opera House 3, *28*, 69

Taylor, Mike 73–74, 93
Tendle, Ashok 92
Terry, Joe 184
Thatcher, Margaret 85, 135–*136*
Thomas, David 29
Thompson, Beatrix Mary Cooper *6*, *8*, 10–13, 17–18, 40; sudden death 51; wish you were here 170, 206
Thompson, Frances 71, *120*–*121*
Thompson, Henry Cooper *6*, 7–*8*, 10
Thompson, Jude 70–*71*, *121*, *139*, *224*
Thompson Milly 219, *222*–*223*
Thompson, Nicky 3–4, 34–35, *40*–*41*, 42–*45*, 57–58, *62*, 64, 66–*71*, *81*, 93–*94*, 102–*103*, 104, 116–*117*, *118*, 119–*122*, 123, 128, 129–131, 139–140, 170, 187, 212–213, 218–219, *221*, *223*–*224*
Thompson, Nim 44–45, 50, 57–59, 66, 70–*71*, 86–87, 119–*122*, 123, 129–130, 139, 170, 187, *223*–*224*
Thompson, Peter Cooper 6, 7–*8*, 10–13, 17–20, 40, 51, 70–*71*: shot dead in Johannesburg 120–121; wedding of daughter Jude *139*–140
Thompson, Phoebe 57, 64, 67, 70–*71*, 87, *102*, 119–*122*, 123, 130, 139, 170, *223*–*224*
Thompson, Sadie *46*–48
Thompson, Sally 71, 120–*121*
Thompson, Sam 52, 57, 66–67,

70–*71*, 87, *102*, 119–*122*, 123, 130, 139, 170, *223*–*224*
Thompson, Sarah 71, *121*
Thompson, Taia 219, *222*–*223*
Thompson, Tonia *224*
Thompson's Eating House, Hungerford, Berkshire *47*–48, 50
Thomson, Alasdair 184
Tray, Lawrence 208
Trevor Estate, The 212–213
Trew, Bob 91
Tulloch, Bruce 199
Tutu, Dr. Desmond 200

UK Task Force for the Federal Republic of Yugoslavia 166
UK Task Force for the Reconstruction of Kosovo 4, 154–155

Viney, Mike 155, 161, 165

Ward, Robert D. 17
Watchorn, Mark 161
Weber, Fleur (later Solomon) *223*
Weber, Lexie *224*
Weber, Monty *224*
Weber, Pip *224*
Weber, Simon *224*
Weeks, John 31, 38
Wellensky, Sir Roy 24
Weston, Michael 126, 128
Weston Water Mill 43–*44*
Whitham, Daniel 207–208
Wigmore Street, London (Debenham and Freebody Department Store) *72*–73
William Nimmo and Partners *98*–101
Willington School, Putney, London 7, 10–*11*
Wixley, Harry *224*
Wixley, John *224*
Wixley, Jude *see* Thompson, Jude
Wixley, Kate *224*
Wordsworth, Stephen 155
Wright, Sir David 154
Wu, Gordon 135

Yangon General Hospital, Myanmar 89–91
York District General Hospital 49
Young, Sir George 134–135

Ziegler, Ron 58
Zimbabwe 141–*142*, *143*–144; *see also* Rhodesia
Zunz, Sir Jack 58, 81–*82*, 84, 85, 104, 129

www.ingramcontent.com/pod-product-compliance
Ingram Content Group UK Ltd.
Pitfield, Milton Keynes, MK11 3LW, UK
UKHW050536150426
5217IPUK00026B/1964